WHY DOES THE WORLD EXIST?

Jim Holt is a longtime contributor to *The New Yorker*—where he has written on string theory, time, infinity, numbers, truth, and bullshit, among other subjects—and the author of *Stop Me If You've Heard This: A History and Philosophy of Jokes*. He is also a frequent contributor to the *New York Times* and the *London Review of Books*. He li... Greenwich Village.

ALSO BY JIM HOLT

Stop Me If You've Heard This: A History and Philosophy of Jokes

WHY DOES THE WORLD EXIST?

An Existential Detective Story

JIM HOLT

P
PROFILE BOOKS

First published in Great Britain in 2012 by
PROFILE BOOKS LTD
3A Exmouth House
Pine Street
London EC1R 0JH
www.profilebooks.com

First published in the United States of America by
W. W. Norton & Company Inc.

1 3 5 7 9 10 8 6 4 2

Excerpt from "Epistemology" from *Ceremony and Other Poems*, copyright 1950
and renewed 1978 by Richard Wilbur, reprinted by permission of Houghton Mifflin
Harcourt Publishing Company. All rights reserved.

Printed and bound in Great Britain by
Clays, Bungay, Suffolk

A CIP catalogue record for this book is available from the British Library.

ISBN 978 1 84668 244 5
eISBN 978 1 84765 410 6

The paper this book is printed on is certified by the © 1996 Forest Stewardship
Council A.C. (FSC). It is ancient-forest friendly. The printer holds FSC chain of
custody SGS-COC-2061

FSC
Mixed Sources
Product group from well-managed
forests and other controlled sources

Cert no. SGS-COC-2061
www.fsc.org
© 1996 Forest Stewardship Council

Contents

WHY DOES THE WORLD EXIST?

Prologue

A Quick Proof That There Must Be Something Rather Than Nothing, for Modern People Who Lead Busy Lives

Suppose there were nothing. Then there would be no laws; for laws, after all, are something. If there were no laws, then everything would be permitted. If everything were permitted, then nothing would be forbidden. So if there were nothing, nothing would be forbidden. Thus nothing is self-forbidding.

Therefore, there must be something. QED.

CONFRONTING THE MYSTERY

And this grey spirit, yearning in desire
To follow knowledge like a sinking star
Beyond the utmost bound of human thought.
—ALFRED, LORD TENNYSON, "*Ulysses*"

I would earnestly warn you against trying to find out the reason for
and explanation of everything. . . . To try and find out the reason for
everything is very dangerous and leads to nothing but
disappointment and dissatisfaction, unsettling your mind and in the
end making you miserable.
—QUEEN VICTORIA, *in a letter to her granddaughter Princess Victoria
of Hesse, 22 August 1883*

. . . well who was the first person in the universe before there was
anybody that made it all who ah that they dont know neither do I . . .
—MOLLY'S SOLILOQUY, *in James Joyce's* Ulysses

I vividly remember when the mystery of existence first swam into my
ken. It was in the early 1970s. I was a callow and would-be rebellious high-school student in rural Virginia. As callow and would-be
rebellious high-school students sometimes do, I had begun to develop
an interest in existentialism, a philosophy that seemed to hold out hope
for resolving my adolescent insecurities, or at least elevating them to a
grander plain. One day I went to the local college library and checked
out some impressive-looking tomes: Sartre's *Being and Nothingness* and

Heidegger's *Introduction to Metaphysics*. It was in the opening pages of the latter book, with its promising title, that I was first confronted by the question *Why is there something rather than nothing at all?* I can still recall being bowled over by its starkness, its purity, its sheer power. Here was the super-ultimate *why* question, the one that loomed behind all the others that mankind had ever asked. Where, I wondered, had it been all my (admittedly brief) intellectual life?

It has been said that the question *Why is there something rather than nothing?* is so profound that it would occur only to a metaphysician, yet so simple that it would occur only to a child. I was too young then to be a metaphysician. But why had I missed the question as a child? In retrospect, the answer was obvious. My natural metaphysical curiosity had been stifled by my religious upbringing. From my earliest childhood I had been told—by my mother and father, by the nuns who taught me in elementary school, by the Franciscan monks at the monastery over the hill from where we lived—that God created the world, and that He created it out of nothing at all. That's why the world existed. That's why I existed. As to why God himself existed, this was left a little vague. Unlike the finite world that He freely created, God was eternal. He was also all-powerful and possessed of every other perfection to an infinite degree. So perhaps He didn't need an explanation for his own existence. Being omnipotent, He might have bootstrapped Himself into existence. He was, to use the Latin phrase, *causa sui*.

This was the story that was imparted to me as a child. It is a story still believed by the vast majority of Americans. For these believers, there is no such thing as the "mystery of existence." If you ask them why the universe exists, they'll say it exists because God made it. If you then ask them why God exists, the answer you get will depend on how theologically sophisticated they are. They might say that God is self-caused, that He is the ground of His own being, that His existence is contained in His very essence. Or they might tell you that people who ask such impious questions will burn in hell.

But suppose you ask nonbelievers to explain why there is a world rather than nothing at all. Chances are, they will not give you a very satisfying answer. In the current "God Wars," those who defend religious belief are wont to use the mystery of existence as a cudgel to beat their neo-atheist opponents with. Richard Dawkins, the evolutionary biologist and professional atheist, is weary of hearing about this supposed mystery. "Time and again," Dawkins says, "my theologian friends return to the point that there had to be a reason why there is something rather than nothing." Christopher Hitchens, another tireless proselytizer for atheism, is often confronted by his opponents with the same question. "If you don't accept that there's a God, how can you explain why the world exists at all?" one slightly thuggish right-wing TV host asked Hitchens, with a note of triumph in his voice. Another such host, this one of the leggy blonde variety, echoed the same religious talking point. "Where did the universe come from?" she demanded of Hitchens. "The idea that this all came out of nothing—that seems to defy logic and reason. What came before the Big Bang?" To which Hitchens replied, "I'd *love* to know what came before the Big Bang."

What options do you have for resolving the mystery of existence once you let go of the God hypothesis? Well, you might expect that science will someday explain not only how the world is, but *why* is it. This, at least, is the hope of Dawkins, who looks to theoretical physics for an answer. "Maybe the 'inflation' that physicists postulate as occupying some fraction of the first yoctosecond of the universe's existence will turn out, when it is better understood, to be a cosmological crane to stand alongside Darwin's biological one," Dawkins has written.

Stephen Hawking, who is actually a practicing cosmologist, takes a different approach. Hawking came up with a theoretical model in which the universe, though finite in time, is completely self-contained, without beginning or end. In this "no-boundary" model, he argued, there is no need for a creator, divine or otherwise. Yet even Hawking doubts that his set of equations can yield a complete resolution to the mystery of existence. "What is it that breathes fire into the equations and makes a universe for them to describe?" he plaintively asks. "Why does the universe go through all the bother of existing?"

The problem with the science option would seem to be this. The universe comprises everything that physically exists. A scientific explanation must involve some sort of physical cause. But any physical cause is by definition part of the universe to be explained. Thus any purely scientific explanation of the existence of the universe is doomed to be circular. Even if it starts from something very minimal—a cosmic egg, a tiny bit of quantum vacuum, a singularity—it still starts with *something*, not *nothing*. Science may be able to trace how the current universe evolved from an earlier state of physical reality, even following the process back as far as the Big Bang. But ultimately science hits a wall. It can't account for the origin of the primal physical state out of nothing. That, at least, is what diehard defenders of the God hypothesis insist.

Historically, when science has seemed incapable of explaining some natural phenomenon, religious believers have been quick to invoke a Divine Artificer to fill the gap—only to be embarrassed when science finally succeeds in filling it after all. Newton, for example, thought that God was needed to make little adjustments from time to time in the orbits of the planets to keep them from colliding. But a century later, Laplace proved that physics was quite capable of accounting for the stability of the solar system. (When Napoléon asked Laplace where God was in his celestial scheme, Laplace famously replied, "*Je n'avais pas besoin de cette hypothèse.*") More recently, religious believers have maintained that blind natural selection alone cannot explain the emergence of complex organisms, so God must be "guiding" the evolutionary process—a contention decisively (and gleefully) refuted by Dawkins and other Darwinians.

Such "God of the gaps" arguments, when they concern the minutiae of biology or astrophysics, tend to blow up in the faces of the religious believers who deploy them. But those believers feel themselves to be on safer ground with the question *Why is there something rather than nothing?* "No scientific theory, it seems, can bridge the gulf between absolute nothingness and a full-fledged universe," the scientifically inclined religious apologist Roy Abraham Varghese has written. "This ultimate origin question is a metascientific question—one which science can ask but not answer." The distinguished Harvard University astronomer (and devout Mennonite) Owen Gingerich agrees. In a lecture titled "God's Universe," delivered at

Harvard's Memorial Church in 2005, Gingerich pronounced the ultimate *why* question to be a "teleological" one—"not for science to grapple with."

Faced with this line of argument, the atheist typically shrugs his shoulders and says the world "just is." Perhaps it exists because it's always existed. Or perhaps it popped into being with no cause at all. In either case, its existence is just a "brute fact."

The brute-fact view denies that the universe as a whole requires any explanation for its existence. It thus avoids the need to posit some sort of transcendental reality, like God, to answer the question *Why is there something rather than nothing?* Yet, intellectually, this feels like throwing in the towel. It's one thing to reconcile yourself to a universe with no purpose and no meaning—we've all done that on a dark night of the soul. But a universe without an explanation? That seems an absurdity too far, at least to a reason-seeking species like ourselves. Whether we realize it or not, we instinctively hew to what the seventeenth-century philosopher Leibniz called the Principle of Sufficient Reason. This principle says, in effect, that explanation goes all the way up and all the way down. For every truth, there must be a reason why it is so and not otherwise; and for every thing, there must be a reason for that thing's existence. Leibniz's principle has been mocked by some as a mere "metaphysician's demand." But it is a bedrock principle of science, where it has been notably successful—so successful, indeed, that one might say it is true on pragmatic grounds: it *works*. The principle seems to inhere in reason itself, since any attempt to argue for or against it already presupposes its validity. And if the Principle of Sufficient Reason is valid, there must be an explanation for the existence of the world, whether we can find it or not.

A world that existed for no reason at all—an irrational, accidental, "just there" world—would be an unnerving one to live in. So, at least, claimed the American philosopher Arthur Lovejoy. In one of his 1933 lectures at Harvard on the "Great Chain of Being," Lovejoy declared that such a world "would have no stability or trustworthiness; uncertainty would infect the whole; anything (except, perhaps, the self-contradictory) might exist and anything might happen, and no one thing would be in itself even more probable than any other."

Are we then doomed to choose between God and the deep brute Absurd?

This dilemma has lurked in the suburbs of my mind ever since I first hit upon the mystery of being. And it has moved me to ponder just what "being" amounts to. The philosopher's term for the ultimate constituents of reality is "substance." For Descartes, the world consisted of two kinds of substance: matter, which he defined as *res extensa* ("extended substance"), and mind, which he defined as *res cogitans* ("thinking substance"). Today, we have pretty much inherited this Cartesian outlook. The universe contains physical stuff: Earth, stars, galaxies, radiation, "dark matter," "dark energy," and so forth. It also contains biological life, which, science has revealed, is physical in nature. In addition, the universe contains consciousness. It contains subjective mental states like joy and misery, the experience of redness, the feel of a stubbed toe. (Are these subjective states reducible to objective physical processes? The philosophical verdict is still out on that question.) An explanation is just a causal story involving items from one or the other of these ontological categories. The impact of the bowling ball caused the pins to drop. Fear of a financial crisis caused a stock market sell-off.

If that's all there is to reality—matter-stuff and mind-stuff, with a web of causal relations between them—then the mystery of being looks hopeless indeed. But perhaps this dualistic ontology is too impoverished. I myself began to suspect as much when, following my teenage flirtation with existentialism, I became infatuated with pure mathematics. The sort of entities mathematicians spend their days pondering—not just numbers and circles, but *n*-dimensional manifolds and Galois systems and crystalline cohomologies—are nowhere found within the realm of space and time. They're clearly not material things. Nor do they seem to be mental. There is no way, for example, that the finite mind of a mathematician could contain an infinity of numbers. Then do mathematical entities really exist? Well, that depends on what you mean by "existence." Plato certainly thought they existed. In fact, he held that mathematical objects, being timeless and unchanging, were more real than the world of things we perceive with our senses. The same was true, he held, of abstract ideas like Goodness and

Beauty. To Plato, such "Forms" constituted genuine reality. Everything else was mere appearance.

We might not want to go that far in revising our notion of reality. Good- ness, Beauty, mathematical entities, logical laws: these are not quite *some- thing*, the way mind-stuff and matter-stuff are. Yet they are not exactly *nothing* either. Might they somehow play a role in explaining *why* there is something rather than nothing?

Admittedly, abstract ideas can't figure in our usual causal explanations. It would be nonsense to say, for example, that Goodness "caused" the Big Bang. But not all explanations have to take this cause-and-effect form; think, for instance, of explaining the point of a chess move. To explain something is, fundamentally, to make it intelligible or understandable. When an explanation is successful, we "feel the key turn in the lock," in the happy phrase of the American philosopher C. S. Peirce. There are many different kinds of explanations, and each one involves a different sense of "cause." Aristotle, for instance, identified four different kinds of cause that might be cited to explain physical occurrences, only one of which (the "efficient" cause) corresponds to our own narrow scientific notion. The most extravagant species of cause in the Aristotelian scheme is the "final" cause—the end or purpose for which something is produced.

Final causes often figure in very bad explanations. (Why does it rain in the spring? So the crops will grow!) Such "teleological" explanations were parodied by Voltaire in *Candide*, and they have been justly rejected by modern science as a way of accounting for natural phenomena. When it comes to accounting for existence as a whole, though, should they be automatically ruled out of court? The assumption that explanations must always involve "things" has been called by one prominent contemporary philosopher, Nicholas Rescher, "a prejudice as deep-rooted as any in West- ern philosophy." Obviously, to explain a given *fact*—such as the fact that there is a world at all—one has to cite other facts. But it doesn't follow that the existence of a given *thing* can be explained only by invoking other things. Maybe a reason for the world's existence should be sought else- where, in the realm of such "un-things" as mathematical entities, objective values, logical laws, or Heisenberg's uncertainty principle. Maybe something along the lines of a teleological explanation might furnish at least a hint as to how the mystery of the world's existence could be resolved.

In the very first philosophy course I took as an undergraduate at the University of Virginia, the professor—a distinguished quondam Oxonian with the evocative name of A. D. Woozley—had us read David Hume's *Dialogues Concerning Natural Religion*. In these dialogues, a trio of fictitious characters—Cleanthes, Demea, and Philo—debate various arguments for the existence of God. Demea, the most religiously orthodox of the three, defends the "cosmological argument," which says, in essence, that the world's existence can be explained only by positing a necessarily existent deity as its cause. In response, the skeptical Philo—who comes closest to being a stand-in for Hume himself—comes up with a seductive bit of reasoning. Although the world seems to be in need of a God-like cause of its existence, Philo observes, that might be due to our own intellectual blindness. Consider, Philo says, the following arithmetical curiosity. If you take any multiple of 9 (like 18, 27, 36, etc.) and add up the digits (1 + 8, 2 + 7, 3 + 6, etc.), you always get 9 back again. To the mathematically naive, this might appear a matter of chance. To the skillful algebraist, by contrast, it is immediately seen to be a matter of necessity. "Is it not probable," Philo then asks, "that the whole economy of the universe is conducted by a like necessity, though no human algebra can furnish a key which solves the difficulty?"

I found this idea of a hidden cosmic algebra—an algebra of being!—irresistible. The very phrase seemed to expand the range of possible explanations for the world's existence. Perhaps the choice was not God versus Brute Fact after all. Perhaps there was a nontheistic explanation for the world's existence—one discoverable by human reason. Although such an explanation wouldn't need to posit a deity, it wouldn't necessarily rule one out either. Indeed, it might even imply the existence of some kind of supernatural intelligence, and in doing so furnish an answer to the precocious child's dread question: "But mommy, who made God?"

How close are we to discovering such an algebra of being? The novelist Martin Amis was once asked by Bill Moyers in a television interview how he thought the universe might have popped into existence. "I'd say we're at least five Einsteins away from answering that question," Amis replied. His estimate seemed about right to me. But, I wondered, could any of those Einsteins be around today? It was obviously not my place to aspire to be one of them. But if I could find one, or maybe two or three or even four of them, and then sort of arrange them in the right order . . . well, that would be an excellent quest.

So that is what I set out to do. My quest to find the beginnings of an answer to the question *Why is there something rather than nothing?* has had many promising leads. Some failed to pan out. Once, for instance, I called a theoretical cosmologist I knew, one noted for his brilliant speculations. I got his voice mail and said that I had a question for him. He called back and left a message on my answering machine. "Leave your question on my voice mail and I'll leave the answer on your machine," he said. This was alluring. I complied. When I returned to my apartment late that evening, the little light was blinking on my answering machine. With some trepidation, I pushed the playback button. "Okay," the cosmologist's recorded voice began, "what you're really talking about is a violation of matter / antimatter parity . . ."

On another occasion, I sought out a certain well-known professor of philosophical theology. I asked him if the existence of the world could be explained by postulating a divine entity whose essence contained his existence. "Are you kidding?" he said. "God is so perfect He doesn't *have* to exist!"

On still another occasion, on a street in Greenwich Village, I ran into a Zen Buddhist scholar I'd been introduced to at a cocktail party. He was said to be an authority on cosmic matters. After a little small talk, I asked him—perhaps, in retrospect, precipitately—"Why is there something rather than nothing?" In response, he tried to bop me on the head. He must have thought it was a Zen kōan.

In searching for enlightenment on the riddle of being, I cast my net fairly wide, talking to philosophers, theologians, particle physicists, cosmologists, mystics, and one very great American novelist. Above all, I looked for versatile and wide-ranging intellects. To have anything really profitable to say about why the world might exist, a thinker must possess more than one kind of intellectual sophistication. Suppose, for example, a scientist has some philosophical acumen. Then he or she might see that the "nothingness" philosophers talked about was conceptually equivalent to something scientifically definable—say, a closed four-dimensional space-time manifold of vanishing radius. By feeding a mathematical description of this null reality into the equations of quantum field theory, one might be able to prove that a small patch of "false vacuum" had a nonzero probability of spontaneously appearing—and that this bit of vacuum, through the marvelous mechanism of "chaotic inflation," would be sufficient to get a full-fledged universe going. If the scientist was also versed in theology, he or she might see how this cosmogonic event could be construed as a backward-in-time emanation from a future "Omega point" that has some of the properties traditionally ascribed to the Judeo-Christian deity. And so forth.

Engaging in such speculative flights takes a good deal of intellectual brio. And brio was amply on display in most of my encounters. One of the pleasures of talking to original thinkers about a matter as profound as the mystery of being is that you get to hear them think out loud. Sometimes they would say the most astonishing things. It was as though I was privileged to peer into their thought processes. This was a cause for awe. But I also found it oddly empowering. When you listen to such thinkers feel their way around the question of why there is a world at all, you begin to realize that your own thoughts on the matter are not quite so nugatory as you had imagined. No one can confidently claim intellectual superiority in the face of the mystery of existence. For, as William James observed, "All of us are beggars here."

Interlude

Could Our World Have Been Created by a Hacker?

W here did our universe come from? Doesn't its sheer existence point to an ultimate creative force at play? This question, when posed by a religious believer to an atheist, generally elicits one of two responses. First, the atheist might say, if you do postulate such a "creative force," you'd better be prepared to postulate another one to explain its existence, and then another one behind that, and so forth. In other words, you end up in an infinite regress. The second atheist response is to say that even if there were an ultimate creative force, there is no reason to think of it as God-like. Why should the First Cause be an infinitely wise and good being, let alone one that is minutely concerned with our inner thoughts and sex lives? Why should it even have a mind?

The idea that our cosmos was somehow "made" by an intelligent being might seem to be a primitive one, if not downright nutty. But before dismissing it entirely, I thought it would be interesting to consult Andrei Linde, who has done more than any other scientist to explain how our cosmos got going. Linde is a Russian physicist who immigrated to the United States in 1990 and who now teaches at Stanford University. While still a young man in Moscow, he came up with a novel theory of the Big Bang that answered three vexing questions: What banged? Why did it bang? And what was going on before it banged? Linde's theory, called "chaotic inflation," explained the overall shape of space and the formation of galaxies.

It also predicted the exact pattern of background radiation left over from the Big Bang that the COBE satellite observed in the 1990s.

Among the curious implications of Linde's theory, one of the most striking is that *it doesn't take all that much to create a universe.* Resources on a cosmic scale are not required, nor are supernatural powers. It might even be possible for someone in a civilization not much more advanced than ours to cook up a new universe in a laboratory. Which leads to an arresting thought: Could that be how *our* universe came into being?

Linde is a handsome, heavy-set man with a full head of silver hair. Among his colleagues he is legendary for his ability to perform acrobatics and baffling sleights of hand, even while a little squiffy.

"When I invented the theory of chaotic inflation, I found that the only thing you needed to get a universe like ours started is a hundred-thousandth of a gram of matter," Linde told me in his Russian-accented English. "That's enough to create a small chunk of vacuum that blows up into the billions and billions of galaxies we see around us. It looks like cheating, but that's how the inflation theory works—all the matter in the universe gets created from the negative energy of the gravitational field. So what's to stop us from creating a universe in a lab? We would be like gods!"

Linde, it should be said, is known for his puckishly gloomy manner, and the preceding words were laced with irony. But he assured me that this cosmogenesis-on-a-lab-bench scenario was feasible, at least in principle.

"There are some gaps in my proof," he conceded. "But what I have shown—and Alan Guth [a codeveloper of inflation theory] and others who have looked at this matter have come to the same conclusion—is that we can't rule out the possibility that our own universe was created by someone in another universe who just felt like doing it."

It struck me that there was a hitch in this scheme. If you started a Big Bang in a lab, wouldn't the baby universe you created expand into your own world, killing people and crushing buildings and so forth?

Linde assured me that there was no such danger. "The new universe would expand into itself," he said. "Its space would be so curved that it would look as tiny as an elementary particle to its creator. In fact, it might end up disappearing from his own world altogether."

But why bother making a universe if it's going to slip away from you, the way Eurydice slipped from the grasp of Orpheus? Wouldn't you want to have some quasi-divine power over how your creation unfolded, some way of monitoring it and making sure the creatures that evolved therein turned out well? Linde's creator seemed very much like the deist concept of God favored by Voltaire and America's founding fathers—a being who set our universe in motion but then took no further interest in it or its creatures.

"You've got a point," Linde said, emitting a slight snuffle of amusement. "At first I imagined that the creator might be able to send information into the new universe—to teach its creatures how to behave, to help them discover what the laws of nature are, and so forth. Then I started thinking. The inflation theory says that a baby universe blows up like a balloon in the tiniest fraction of a second. Suppose the creator tried to write something on the surface of the balloon, like "PLEASE REMEMBER THAT I MADE YOU." The inflationary expansion would make this message exponentially huge. The creatures in the new universe, living in a tiny corner of one letter, would never be able to read the whole message."

But then Linde thought of another channel of communication between creator and creation—the only one possible, as far as he could tell. The creator, by manipulating the cosmic seed in the right way, would have the power to ordain certain physical parameters of the universe he ushers into being. He could determine, for example, what the numerical ratio of the electron's mass to the proton's will be. Such numbers, called the constants of nature, look utterly arbitrary to us: there is no apparent reason why they should take the value they do rather than some other value. (Why, for instance, is the strength of gravity in our universe determined by a number with the digits "6673"?) But the creator, by fixing certain values for these constants, could write a subtle message into the very structure of the universe. And, as Linde pointed out with evident relish, such a message would be legible only to physicists.

Was he joking?

"You might take this as a joke," he said. "But perhaps it is not entirely absurd. It may furnish the explanation for why the world we live in is so weird, so far from perfect. On the evidence, our universe wasn't created by a divine being. It was created by a physicist hacker!"

From a philosophical perspective, Linde's little story underscores the danger of assuming that the creative force behind our universe, if there is one, must correspond to the traditional image of God: omnipotent, omniscient, infinitely benevolent, and so on. Even if the cause of our universe is an intelligent being, it could well be a painfully incompetent and fallible one, the kind that might flub the cosmogenic task by producing a thoroughly mediocre creation. Of course, orthodox believers can always respond to a scenario like Linde's by saying, "Okay, but who created the physicist hacker?" Let's hope it's not hackers all the way up.

2

PHILOSOPHICAL TOUR D'HORIZON

The riddle does not exist.

—LUDWIG WITTGENSTEIN, Tractatus Logico-Philosophicus,

proposition 6.5

The crux of the mystery of existence, as I have said, is summed up in the question *Why is there something rather than nothing?* William James called this question "the darkest in all philosophy." The British astrophysicist Sir Bernard Lovell observed that pondering it could "tear the individual's mind asunder." (Indeed, psychiatric patients have been known to be obsessed by it.) Arthur Lovejoy, who founded the academic field known as the History of Ideas, observed that the attempt to answer it "constitutes one of the most grandiose enterprises of the human intellect." Like all deep incomprehensibilities, it lends an opening to jocularity. Some decades ago, when I put the question to the American philosopher Arthur Danto, he replied, with mock irritability, "Who says there's not nothing?" (As will soon become apparent, this response is not entirely a joke.) A still better answer was supplied by Sidney Morgenbesser, late Columbia University philosopher and legendary wag. "Professor Morgenbesser, why is there something rather than nothing?" a student asked him one day. To which Morgenbesser replied, "Oh, even if there was nothing, you still wouldn't be satisfied!"

But the question cannot be laughed away. Each of us, as Martin Heidegger observed, is "grazed by its hidden power":

The question looms in moments of great despair when things tend to lose all their weight and all meaning becomes obscured. It is present in moments of rejoicing, when all the things around us are transfigured and seem to be there for the first time. . . . The question is upon us in boredom, when we are equally removed from despair and joy, and everything about us seems so hopelessly commonplace that we no longer care whether anything is or is not.

Ignoring this question is a symptom of mental deficiency—so, at least, claimed the philosopher Arthur Schopenhauer. "The lower a man is in an intellectual respect, the less puzzling and mysterious existence itself is to him," Schopenhauer wrote. What raises man above other creatures is that he is conscious of his finitude; the prospect of death brings with it the conceivability of nothingness, the shock of nonbeing. If my own self, the microcosm, is ontologically precarious, so perhaps is the macrocosm, the universe as a whole. Conceptually, the question *Why does the world exist?* rhymes with the question *Why do I exist?* These are, as John Updike observed, the two great existential mysteries. And if you happen to be a solipsist—that is, if you believe, as did the early Wittgenstein, "I am my world"—the two mysteries fuse into one.

FOR A QUESTION that is supposed to be timeless and universal, it is strange that nobody explicitly asked, *Why is there something rather than nothing?* until the modern era. Perhaps it's the "nothing" part of the question that makes it truly modern. Premodern cultures have their creation myths to explain the origin of the universe, but such myths never start from sheer nothingness. They always presuppose some primordial beings or stuff out of which reality arose. In a Norse myth current around 1200 CE, for instance, the world began when a primeval region of fire melted a primeval region of frost, giving rise to liquid drops that quickened into life and took the form of a wise giant called Ymer and a cow called Audhumla— whence eventually sprang the rest of existence as the Vikings knew it. According to a somewhat more economical creation myth, that of the

African Bantus, the entire contents of the universe—sun, stars, land, sea, animals, fish, mankind—are literally vomited out of the mouth of a nauseated being called Bumba. Cultures that have no creation myth to explain how the world came into being are rare, but not unknown. One such is the Pirahã, an amusingly perverse Amazon tribe. When anthropologists ask Pirahã tribespeople what preceded the world, they invariably reply, "It has always been this way."

A theory about the birth of the universe is called a cosmogony, from the Greek *kosmos*, meaning "universe," and *gonos*, meaning "produce" (the same as the root for "gonad"). The ancient Greeks were the pioneers of rational cosmogony, as opposed to the mythopoetic variety exemplified by creation myths. Yet the Greeks never raised the question of why there is a world rather than *nothing at all*. Their cosmogonies always involved some sort of starting material, usually rather messy. The natural world, they held, came into existence when order was imposed on this primal mess: when Chaos became Cosmos. (It is interesting that the words "cosmos" and "cosmetic" have the same root, the Greek word for "adornment" or "arrangement.") As to what this original Chaos might have been, the Greek philosophers had various guesses. For Thales, it was watery, a kind of ur-Ocean. For Heraclitus, it was fire. For Anaximander, it was something more abstract, an indeterminate material called "the Boundless." For Plato and Aristotle, it was a formless substrate that might be taken as a prescientific notion of space. The Greeks did not worry too much about where this ur-matter came from. It was simply assumed to be eternal. Whatever it was, it was certainly not *nothing*—the very idea of which was inconceivable to the Greeks.

Nothingness was alien to the Abrahamic tradition too. The book of Genesis has God creating the world not out of nothing, but out of a chaos of earth and water "without form and void"—*tohu bohu*, in the original Hebrew. Early in the Christian era, however, a new way of thinking began to take hold. The notion that God needed some sort of stuff to fashion a world seemed to put a limit on his presumably infinite creative powers. So, around the second or third century CE, the church fathers adduced a radical new cosmogony. The world, they proclaimed, was summoned into existence by God's creative word alone, without any preexisting material to make it out of. This doctrine of creation *ex nihilo* later became part of Islamic theology, figuring in the kalām argument for the existence of God. It also entered medieval Jewish thought. In his reading of the opening passage of Genesis, the Jewish philosopher Maimonides affirmed that God created the world out of nothing.

To say God created the world "out of nothing" is not to elevate nothingness into an entity, on par with the divine. It merely means that God didn't create the world out of *anything*. So insisted Saint Thomas Aquinas, among other Christian theologians. Still, the doctrine of creation *ex nihilo* appeared to sanction the idea of nothingness as a genuine ontological possibility. It made it conceptually possible to ask why there is a world rather than *nothing at all*.

And a few centuries later, someone finally did—a foppish and conniving German courtier who also ranks among the greatest intellects of all time: Gottfried Wilhelm Leibniz. The year was 1714. Leibniz, then sixty-eight, was nearing the end of a long and absurdly productive career. He had, at the same time as Newton and quite independently, invented the calculus. He had single-handedly revolutionized the science of logic. He had created a fantastic metaphysics based on an infinity of soul-like units called "monads," and on the axiom—later cruelly mocked by Voltaire in *Candide*—that this is "the best of all possible worlds." Despite his fame as a philosopher-scientist, Leibniz was left behind in Hanover when his royal employer, the elector Georg Ludwig, went to Britain to become the newly crowned King George I. Leibniz was in declining health; within two years he would be dead, expiring (according to his secretary) with the release from his body of a great cloud of noxious gas.

It was in these gloomy circumstances that Leibniz produced his final philosophical writings, among them an essay titled "Principles of Nature and Grace, Based on Reason." In this essay, he put forth what he called the "Principle of Sufficient Reason," which says, in essence, that there is an explanation for every fact, an answer for every question. "This principle having been stated," Leibniz wrote, "the first question which we have a right to ask will be, 'Why is there something rather than nothing?'"

For Leibniz, the ostensible answer was easy. For reasons of career advancement, he had always pretended to hew to religious orthodoxy. The reason for the world's existence, he accordingly claimed, was God, who created it through his own free choice, motivated by his infinite goodness.

But what was the explanation for God's own existence? Leibniz had an answer to this question too. Unlike the universe, which exists contingently, God is a *necessary* being. He contains within Himself the reason for His own existence. His nonexistence is logically impossible.

Thus, no sooner was the question *Why is there something rather than nothing?* raised than it was dispatched. The universe exists because of God. And God exists because of God. The Godhead alone, Leibniz declared, can furnish the ultimate resolution to the mystery of existence.

But the Leibnizian resolution to the mystery of existence did not prevail for long. In the eighteenth century, both David Hume and Immanuel Kant—philosophers who were at loggerheads on most issues—attacked the notion of "necessary being" as an ontological cheat. There are, to be sure, entities whose existence is logically *impossible*—a square circle, for instance. But no entity's existence, Hume and Kant agreed, is guaranteed as a matter of pure logic. "Whatever we can conceive as existent we can also conceive as non-existent," Hume wrote. "There is no being, therefore, whose non-existence implies a contradiction"—including God.

But if God does not exist necessarily, then a wholly novel metaphysical possibility presents itself: the possibility of *absolute nothingness*—no world, no God, no anything. Oddly, however, neither Hume nor Kant took the question *Why is there something rather than nothing?* seriously. For Hume, any proposed answer to this question would be "mere sophistry and illusion," since it could never be grounded in our experience. For Kant, attempting to explain the whole of being would perforce involve an illegitimate extension of the concepts that we use to structure the world of our experience—concepts like *causality* and *time*—to a reality transcending this world, the reality of "things in themselves." The result, Kant held, could be only error and inconsistency.

Chastened, perhaps, by such Humean and Kantian strictures, subsequent philosophers largely shied away from confronting the question *Why is there something rather than nothing?* The great pessimist Schopenhauer, who declared the mystery of existence to be "the balance wheel which maintains in motion the watch of metaphysics," nevertheless called those who pretended to resolve it "fools," "vain boasters," and "charlatans." The German romantic Friedrich Schelling stated that "the main function of all philosophy is the solution of the problem of the existence of the world." Yet Schelling soon decided that it was impossible to give a rational account of existence; the most we could say, he felt, was that the world arose out of the abyss of eternal nothingness by an incomprehensible leap. Hegel wrote a good deal of obscure prose about "the vanishing of being into nothing and the vanishing of nothing into being," but his dialectical maneuvers were dismissed by the ironic Danish thinker Søren Kierkegaard as little better than "spice-seller's explanations."

The beginning of the twentieth century saw a modest revival of interest in the mystery of existence, mainly thanks to the French philosopher Henri Bergson. "I want to know why the universe exists," Bergson declared in his 1907 book, *Creative Evolution*. All existence—matter, consciousness, God himself—was, it seemed to Bergson, a "conquest over nothingness." But after much pondering, he concluded that this conquest was not really so miraculous. The whole something-versus-nothing question was based on an illusion, he came to believe: the illusion that it was possible for there to be nothing at all. By a series of dubious arguments, Bergson purported to prove that the idea of absolute nothingness was as self-contradictory as the idea of a round square. Since nothingness was a pseudo-idea, he concluded, the question *Why is there something rather than nothing?* was a pseudo-question.

This killjoy conclusion certainly made no impression on Martin Heidegger, for whom nothingness was all too real, a sort of negating force that menaced the realm of being with annihilation. At the very beginning of a series of lectures delivered in 1935 at the University of Freiburg—where he had been given the job of rector after proclaiming his allegiance to Hitler's national socialism—Heidegger declared "Why is there being rather than nothing at all?" to be the "deepest," "the most far-reaching," and "the most fundamental of all questions."

And what did Heidegger do with this question as the lectures progressed? Not a lot. He dilated on its existential pathos. He dabbled in amateur etymology, piling up Greek, Latin, and Sanskrit words related to *Sein*, the German word for "being." He rhapsodized about the poetic virtues of the pre-Socratics and the Greek tragedians. At the conclusion of the final lecture, Heidegger observed that "being able to ask a question means being able to wait, even one's whole life long"—which must have had those in the audience who had been hoping for a hint of an answer wearily nodding their heads.

Heidegger was, without question, the most influential philosopher of the twentieth century in continental Europe. But in the English-speaking world, it was Ludwig Wittgenstein who had the greatest philosophical sway. Wittgenstein and Heidegger were born in the same year (1889). They were pretty much opposites when it came to character: Wittgenstein was brave and ascetic, Heidegger treacherous and vain. Yet they were equally seduced by the mystery of existence. "It is not *how* things are in the world that is mystical, but *that* it exists," Wittgenstein averred in one of the lapidary numbered propositions—6.44, to be precise—in the sole work he published in his lifetime, *Tractatus Logico-Philosophicus*. Some years earlier, in the notebooks he kept as a soldier in the Austrian army during the First World War, Wittgenstein wrote in the entry of October 26, 1916, "Aesthetically, the miracle is that the world exists." (Later that day, he made the entry, "Life is serious, art is gay"—this while fighting on the Russian front.) Wonder and amazement at the existence of the world was, Wittgenstein said, one of the three experiences that enabled him to fix his mind on ethical value. (The other two were the feeling of being absolutely safe, and the experience of guilt.) Yet, as with all truly important matters—ethical value, the meaning of life and death—attempting to explain the "aesthetic miracle" of the world's existence was futile; it took one beyond the limits of language, Wittgenstein held, into the realm of the unsayable. While he "deeply respected" the urge to ask *Why is there something rather than nothing?* he ultimately believed the question to be senseless. As he starkly put it in *Tractatus* proposition 6.5, "*The riddle* does not exist."

Ineffable though it may have been to Wittgenstein, the mystery of existence nevertheless filled him with awe and gave him a sense of spiritual illumination. For many of the British and American philosophers in his wake, by contrast, it seemed a woolly waste of time. Typifying their dismissive attitude was A. J. "Freddy" Ayer, the British champion of logical positivism, sworn enemy of metaphysics, and self-avowed philosophical heir of David Hume. In a 1949 BBC radio broadcast, Ayer engaged Frederick Copleston, a Jesuit priest and historian of philosophy, in a debate on the existence of God. Much of the Ayer-Copleston debate, as it turned out, was taken up with the question of why there is something rather than nothing. For Father Copleston, this question was an opening to the transcendent, a way of seeing how God's existence is "the ultimate ontological explanation of phenomena." For Ayer, his atheist opponent, it was illogical twaddle.

"Supposing," Ayer said, "you asked a question like 'Where do all things come from?' Now that's a perfectly meaningful question as regards any given event. Asking where it came from is asking for a description of some event prior to it. But if you generalize that question, it becomes meaningless. You're then asking what event is prior to all events. Clearly no event can be prior to all events. Because it's a member of the class of all events it must be included in it, and therefore can't be prior to it."

Wittgenstein, who listened to the radio broadcast, later told a friend that he found Ayer's reasoning to be "incredibly shallow." Still, the debate was deemed so close that a televised rematch was scheduled a few years later. But Ayer and Copleston were plied with so much whiskey while a technical malfunction was being corrected that both men were reduced to incoherence by the time the debate commenced.

The disagreement between Ayer and Copleston on the meaningfulness of the question *Why is there something rather than nothing?* came down to a dispute over the very nature of philosophy. And the vast majority of philosophers, at least in the English-speaking world, sided with Ayer in this dispute. There were two kinds of truths, the orthodoxy went: logical truths and empirical truths. Logical truths depended only on the meanings of words. The necessities they expressed, like *All bachelors are unmarried*, were merely verbal necessities. Hence, logical truths could not explain anything about reality. Empirical truths, by contrast, depended on the evidence furnished by the senses. They were the province of scientific inquiry. And it was generally conceded that the question of why the world exists was beyond the reach of science. A scientific explanation, after all, could account for one bit of reality only in terms of other bits; it could never account for reality *as a whole*. So the existence of the world could be only a brute fact. Bertrand Russell summed up the philosophical consensus: "I should say that the universe is just there, and that is all."

Science, for the most part, concurred. The brute-fact take on existence is a fairly comfortable one if you assume that the universe has *always* been around. And that, indeed, was what most of the greatest scientists of the modern era—including Copernicus, Galileo, and Newton—believed. Einstein was convinced that the universe was not only eternal but also, on the whole, unchanging. So when, in 1917, he applied his general theory of relativity to spacetime as a whole, he was perplexed to find that his equations implied something radically different: the universe must be either expanding or contracting. This struck him as grotesque, so he added a fiddle-factor to his theory so that it would allow for a universe that was both eternal and unchanging.

It was an ordained priest who had the nerve to push relativity to its logical conclusion. In 1927, Georges Lemaître, of the University of Louvain in Belgium, worked out an Einsteinian model of the universe in which space was expanding. Reasoning backward, Father Lemaître proposed that at some definite point in the past the entire universe must have originated from a primeval atom of infinitely concentrated energy. Two years later, Lemaître's expanding-universe model was confirmed by the American astronomer Edwin Hubble, whose observations at the Mount Wilson Observatory in California established that the galaxies everywhere around us were indeed receding. Both theory and empirical evidence pointed to the same verdict: the universe must have had an abrupt beginning in time.

Churchmen rejoiced. Scientific proof of the biblical account of creation had, they believed, dropped into their laps. Pope Pius XII, opening a conference at the Vatican in 1951, declared that this new theory of cosmic origins bore witness "to that primordial *Fiat lux* uttered at the moment when, along with matter, there burst forth from nothing a sea of light and radiation.... Hence, creation took place in time, therefore there is a creator, therefore God exists!"

Those at the other ideological extreme gnashed their teeth—Marxists in particular. Quite aside from its religious aura, the new theory contradicted their belief in the infinity and eternity of matter, which was one of the axioms of Lenin's dialectical materialism. Accordingly, the theory was dismissed as "idealistic." The *Marxisant* physicist David Bohm rebuked the developers of the theory as "scientists who effectively turn traitor to science, and discard scientific facts to reach conclusions that are convenient to the Catholic Church." Atheists of a non-Marxist stripe were also recalcitrant. "Some younger scientists were so upset by these theological trends that they resolved simply to block their cosmological source," commented the German astronomer Otto Heckmann, a prominent investigator of cosmic expansion. The dean of the profession, Sir Arthur Eddington, wrote that "the notion of a beginning is repugnant to me.... I simply do not believe that the present order of things started off with a bang ... the expanding Universe is preposterous ... incredible ... it leaves me cold."

Even some believing scientists were troubled. The cosmologist Sir Fred Hoyle felt that an explosion was an undignified way for the world to begin, rather like "a party girl jumping out of a cake." In a BBC interview in the 1950s, Hoyle sardonically referred to the hypothesized origin as "the Big Bang." The term stuck.

Einstein, not long before his death in 1955, managed to overcome his metaphysical scruples about the Big Bang. He referred to his earlier attempt to dodge it by an ad hoc theoretical device as "the greatest blunder of my career." As for Hoyle and the rest of the skeptics, they were finally won over in 1965 when two scientists at Bell Labs in New Jersey accidently detected a pervasive microwave hiss that turned out to be the echo of the Big Bang. (At first the scientists thought the hiss was caused by pigeon droppings on their antenna.) If you turn on your television and tune it between stations, about 10 percent of that black-and-white speckled static you see is caused by photons left over from the birth of the universe. What greater proof of the reality of the Big Bang—you can watch it on TV.

Whether or not the universe had a creator, the finding that it came into existence at a finite time in the past—13.7 billion years ago, according to the latest cosmological calculations—appeared to make a mockery of the idea that it was ontologically self-sufficient. Anything that exists by its own nature, it seems reasonable to assume, must be eternal and imperishable. The universe now looked to be neither of these things. Just as it winked into existence with an initial Big Bang, expanding and evolving into its present form, so too it might wink out of existence in some distant future epoch with an annihilating Big Crunch. (Whether the ultimate fate of the universe will be a Big Crunch, a Big Chill, or a Big Crack-up is a wide-open question in cosmology today.) The life of the universe, like each of our lives, may be a mere interlude between two nothings.

Thus did the discovery of the Big Bang make the question *Why is there something rather than nothing?* much harder to dodge. "If the universe hadn't always existed, science would be confronted by the need for an explanation of its existence," observed Arno Penzias, who shared a Nobel Prize for detecting the afterglow of the Big Bang. Not only was the original *why* question a live one, but it now needed to be supplemented by a *how* question: *How could something have arisen from nothing?* Besides giving renewed hope to religious apologists, the Big Bang hypothesis opened up a new and purely scientific inquiry into the ultimate origin of the universe. And the explanatory possibilities seemed to multiply. There were, after all, two revolutionary developments in twentieth-century physics. One of them, Einstein's relativity theory, led to the conclusion that the universe had a beginning in time. The other, quantum mechanics, had even more radical implications. It threw into doubt the very idea of cause and effect. According to quantum theory, events at the micro-level happen in aleatory fashion; they violate the classical principle of causation. This opened up the conceptual possibility that the seed of the universe might itself have come into being without a cause, supernatural or otherwise. Perhaps the world arose *spontaneously* from sheer nothingness. All existence might be chalked up to a random fluctuation in the void, a "quantum tunneling" from nothingness into being. Exactly how this could have happened has become the province of a small but influential group of physicists who are sometimes referred to as "nothing theorists." With a mixture of metaphysical chutzpah and naivete, these physicists—who include Stephen Hawking among their number—think they might be able to resolve a mystery heretofore considered untouchable by science.

INSPIRED, PERHAPS, BY this scientific ferment, philosophers have been showing more ontological boldness. Logical positivism, which had dismissed the question *Why is there something rather than nothing?* as nonsensical, was defunct by the 1960s, a victim of its own inability to arrive at a workable distinction between sense and nonsense. In its wake, metaphysics—the project of characterizing reality as a whole—has seen a revival. Even in the Anglo-Saxon world, "analytical" philosophers are no longer embarrassed to grapple with metaphysical issues. The most audacious of the many professional philosophers who have confronted the mystery of existence in the last few decades was Robert Nozick of Harvard University, who died at the age of sixty-three in 2002. Although best known as the author of the libertarian classic *Anarchy, State, and Utopia*, Nozick was obsessed with the question *Why is there something rather than nothing?*, devoting a fifty-page section of his later book *Philosophical Explanations* to the various possibilities for answering it—some of them quite wild. He invited the reader to imagine nothingness as a force "sucking things into non-existence." He posited a "principle of fecundity" that sanctions the simultaneous existence of all possible worlds. He talked of having some kind of mystical insight into reality's foundation. As for his colleagues who might have found his attempts to answer the ultimate question a little strange, Nozick was unapologetic: "Someone who proposes a non-strange answer shows he didn't understand the question."

TODAY, THINKERS REMAIN divided into three camps by the question *Why is there something rather than nothing?* The "optimists" hold that there *has to be* a reason for the world's existence, and that we may well discover it. The "pessimists" believe that there *might be* a reason for the world's existence, but that we'll never know for sure—perhaps because we see too

little of reality to be aware of the reason behind it, or because any such reason must lie beyond the intellectual limits of humans, which were tooled
by nature for survival, not for penetrating the inner nature of the cosmos.
Finally, the "rejectionists" persist in believing that there *can't be* a reason
for the world's existence, and hence that the very question is meaningless.

You don't have to be a philosopher or a scientist to join one of these
camps. Everyone is entitled. Marcel Proust, for instance, seems to have
placed himself among the pessimists. The narrator of his novel *Remem-
brance of Things Past*, musing on how the Dreyfus affair had split French
society into warring factions, observes that political wisdom may be powerless to end the civil strife, just as "in philosophy, pure logic is powerless
to tackle the problem of existence."

But suppose you're an optimist. What is the most promising approach
to the mystery of existence? Is it the traditional theistic approach, which
looks to a God-like entity as the necessary cause and sustainer of all being?
Is it the scientific approach, which draws on ideas from quantum cosmology to explain why a universe was bound to leap into existence out of the
void? Is it a purely philosophical approach, which seeks to deduce a reason
for the world's existence from abstract considerations of value, or from the
sheer impossibility of nothingness? Is it some sort of mystical approach,
which aims to satisfy the craving for a cosmic rationale through direct
illumination?

All of these approaches have their contemporary proponents. All of
them, at first blush, seem worth pursuing. Indeed, it is only by thinking
of the mystery of existence from every available angle that we have any
hope of resolving it. To those who consider the question *Why is there something rather than nothing?* hopelessly elusive or downright incoherent, it
might be pointed out that intellectual progress often consists in the refine-
ment of precisely such questions, in ways unforeseeable to those who first
ask them. Take another question, posed twenty-five hundred years ago
by Thales and his fellow pre-Socratics: *What are things made of?* Asking
a question of such all-encompassing generality might sound naive, even
childish. But, as the Oxford philosopher Timothy Williamson observed,
the pre-Socratic philosophers "were asking one of the best questions ever

to have been asked, a question that has painfully led to much of modern science." To have dismissed it from the outset as unanswerable would have been "a feeble and unnecessary surrender to despair, philistinism, cowardice or indolence."

The mystery of existence, however, might seem uniquely futile among such questions. For, as William James put it, "from nothing to being there is no logical bridge." But can this be known before any attempt is made to construct such a bridge? Other seemingly impossible bridges have been successfully built: from nonlife to life (thanks to molecular biology), from finite to infinite (thanks to the mathematical theory of sets). Today, those working on the problem of consciousness are trying to bridge mind and matter, and those trying to unify physics are trying to bridge matter and mathematics. With such conceptual linkages taking form, one can perhaps begin to see the faint outlines of a bridge between Nothing and Something (or perhaps a tunnel, if the quantum theorists are right). One can only hope it doesn't turn out to be a bridge of asses.

THE MOTIVES FOR pursuing the mystery of existence are not just intellectual ones. They are also emotional. Our emotions typically have objects; they are *about* something. I am sad about the death of my dog. You are overjoyed that the Yankees are in the World Series. Othello is enraged at Desdemona's infidelity. But some emotional states seem to be "free-floating," without any determinate objects. Kierkegaard's dread, for instance, was directed at nothing, or at everything. Moods like depression and exhilaration, if they have any object at all, seem to be about existence itself. Heidegger maintained that at the deepest level this is true of all emotions.

What sort of emotion is appropriate when the object of that emotion is the world as a whole?

This question divides people into two categories: those who smile on existence, and those who frown on it. For a notable frowner, consider Arthur Schopenhauer, whose philosophical pessimism influenced such later thinkers as Tolstoy, Wittgenstein, and Freud. If we are astonished at the existence of the world, Schopenhauer declared, our astonishment is one of *dismay* and *distress*. That is why "philosophy, like the overture to *Don Juan*, starts with a minor chord." We live not in the best of all worlds, he went on, but in the worst. Nonexistence "is not only conceivable, but even preferable to its existence." Why? Well, in Schopenhauer's metaphysics, the entire universe is a great manifestation of striving, one vast will. All of us, with our seemingly individual wills, are merely little bits of this cosmic will. Even inanimate nature—the attractive force of gravity, the impenetrability of matter—partakes in it. And will, for Schopenhauer, is essentially *suffering*: there is no end that, if achieved, would bring contentment; the will is either frustrated and miserable, or sated and bored. Schopenhauer was the first thinker to import this Buddhist strain into Western thought. The only way out of suffering, he taught, is to extinguish the will and thereby enter a state of nirvana—which is as close to nonexistence as we can get: "No will: no idea, no world. Before us there is certainly only nothingness." It must be said that Schopenhauer himself hardly practiced the pessimistic ascetism he preached: he was fond of the pleasures of the table; enjoyed many sensual affairs; was quarrelsome, greedy, and obsessed with his fame. He also kept a poodle named Atma—Sanskrit for "world soul."

In the last century, Schopenhauerian frowners have predominated, at least in the literary world. An especially heavy concentration of them could be found on the boulevards of Paris. Take E. M. Cioran, the Romanian writer who came to Paris and reinvented himself as an existential *flâneur*. Not even the charms of his adopted city could ease his nihilistic despair. "When you have understood that nothing *is*," Cioran wrote, "that things do not even deserve the status of appearances, you no longer need to be saved, you are saved, and miserable forever." Samuel Beckett, another expatriate in Paris, was similarly afflicted by the emptiness of being. Why, Beckett wanted to know, is the cosmos indifferent to us? Why are we such an insignificant part of it? Why is there a world at all?

Jean-Paul Sartre, in his moods, could be similarly jaundiced about existence. Roquentin, the autobiographical hero of Sartre's novel *Nausea*, finds himself "choked with rage" at the "monstrous lumps" of "gross, absurd being" that environ him as he sits under a chestnut tree in the fictional village of Bouville (French for "Mudville"). The sheer contingency of it all strikes him as not just absurd but downright obscene. "You couldn't even wonder where all that sprang from, or how it was that a world came into existence, rather than nothingness," Roquentin muses, whereupon he is moved to shout, "Filth!" at the "tons and tons of existence" and then lapses into an "immense weariness."

American literary figures have tended to wear their ontological pessimism more cheerfully. The playwright Tennessee Williams, for example, simply observed that "a vacuum is a hell of a lot better than some of the stuff nature replaces it with," and then had another whiskey. John Updike channeled his ambivalence about Being into his fictional alter-ego, that blocked, priapic, and despair-prone Jewish novelist Henry Bech. In one Updike story, Bech is invited to give a reading at a Southern girls' college, where he is regarded as a literary star. At a dinner in his honor after the reading, he "looked around the ring of munching females and saw their bodies as a Martian or a mollusc might see them, as pulpy stalks of bundled nerves oddly pinched to a bud of concentration in the head, a hairy bone knob holding some pounds of jelly in which a trillion circuits, mostly dead, kept records, coded motor operations, and generated an excess of electricity that pressed into the hairless side of the head and leaked through the orifices, in the form of pained, hopeful noises and a simian dance of wrinkles." Bech has a nihilistic epiphany: "the void should have been left unvexed, should have been spared this trouble of matter, of life, and, worst, of consciousness." All existence, he declares to himself, is but a "blot on nothingness." Yet, in his sunnier humors—or when he is affecting sunniness during the taping of a literary interview—Updike's Bech is capable of smiling upon Being: "He believed, if this tape recorder must know . . . in the dignity of the inanimate, the intricacy of the animate, the beauty of the average woman, and the common sense of the average man." In short, Bech believed "in the goodness of something vs. nothing." Bech's spasm of ontological optimism puts one in mind of a famous nineteenth-century New England transcendentalist, Margaret Fuller, who was fond of exclaiming, "I accept the universe!" (to which the acidulous Thomas Carlyle responded, "Gad, she'd better").

Perhaps the most ringing endorsement of the goodness of the world is not literary or philosophical, but musical. It is offered by Haydn in his oratorio *The Creation*. At first, all is musical chaos, a mixture of eerie harmonics and fragmentary melodies. Then comes the creative moment, when God declares, "Let there be light!" As the singers respond, "And there was light," both orchestra and chorus mark the miracle by bursting into a powerful and sustained C-major triad—the very opposite of the gloomy Schopenhauer's "minor chord."

The attitude one takes toward existence as a whole shouldn't merely be a matter of temperament—of whether or not one is liverish, or of how well one slept the previous night. It should be subject to rational evaluation. And it is only by exploring the question *Why is there something rather than nothing?* that we might come to see the value of existence in a rational light.

Could it be, for instance, that the world exists precisely because it is, on the whole, better than nothing? There are indeed philosophers who believe such a thing. They call themselves "Axiarchists." (The word comes from the Greek for "value rules!") They think the cosmos may have exploded into being in answer to a need for goodness. If they are right, the world, and our existence within it, may be better than it appears to us. We should be on the lookout for its subtler virtues, like hidden harmonies and dappled things.

Others hold that the triumph of Being over Nothingness may well have been a matter of blind chance. There are, after all, lots of ways for there to be Something—worlds in which everything is blue, worlds made of cream cheese, and so on—but there is only one Nothing. Assuming that all possible realities were assigned equal chances in the cosmic lottery, it is overwhelmingly likely that one of the many Somethings would win, not the lonely Nothing. If this blind-chance view of reality turned out to be right, we would have to revise our attitude toward existence downward a bit. For if reality is the outcome of a cosmic lottery, it is probable that the winning world will be a mediocre one: neither very good nor very wicked, neither very neat nor very messy, neither very beautiful nor very ugly. That is because mediocre possibilities are common, and truly excellent or awful ones rare.

If, on the other hand, the answer to the puzzle of existence turns out to be a theistic or quasi-theistic one—that is, if it involves something like a creator—then the attitude one takes toward the world would depend on the nature of that creator. The major monotheistic religions hold that the world was created by a God that is all-good and all-powerful. If this is true, then one is more or less obliged to regard the world in a favorable light, notwithstanding physical imperfections like redundant elementary particles and imploding stars, and moral imperfections like cancer in children and the Holocaust. But some religions have entertained a different doctrine of creation. The Gnostics, a heretical movement that flourished in the early Christian era, held that the material world was created not by a benevolent deity, but by an evil demiurge. Thus they deemed themselves justified in loathing material reality. (A useful compromise between the Christians and the Gnostics might be my own position: that the universe was created by a being that is 100 percent malevolent but only 80 percent effective.)

Of all the possible resolutions to the mystery of existence, perhaps the most exhilarating would be the discovery that, contrary to all appearances, the world is *causa sui*: the cause of itself. This possibility was first raised by Spinoza, who boldly (if a little obscurely) reasoned that all reality consists of a single infinite substance. Individual things, both physical and mental, are merely temporary modifications of this substance, like waves on the surface of the sea. Spinoza referred to this infinite substance as *Deus sive Natura*: "God or Nature." God could not possibly stand apart from nature, he reasoned, because then each would limit the other's being. So the world itself is divine: eternal, infinite, and the cause of its own existence. Hence, it is worthy of our awe and reverence. Metaphysical understanding thus leads to "intellectual love" of reality—the highest end for humans, according to Spinoza, and the closest we can come to immortality.

Spinoza's picture of the world as *causa sui* captivated Albert Einstein. In 1921, a New York rabbi asked Einstein if he believed in God. "I believe in Spinoza's God," he answered, "who reveals himself in the orderly harmony of what exists, not in a God who concerns himself with the fates and actions of human beings." The idea that the world somehow holds the key to its own existence—and hence that it exists necessarily, not as an accident—jibes with the thinking of some metaphysically inclined physicists, such as

Sir Roger Penrose and the late John Archibald Wheeler (who coined the term *black hole*). It has even been conjectured that the human mind plays a critical role in the self-causing mechanism. Although we seem to be a negligible part of the cosmos, it is our consciousness that gives reality to it as a whole. On this picture, sometimes called the "participatory universe," reality is a self-sustaining causal loop: the world creates us, and we in turn create the world. It's a bit like Proust's great work, which records the progress and the sufferings of its hero through thousands of pages until, at the end, he resolves to write the very novel we have been reading.

Such a Promethean fantasy—we are the world's author as well as its plaything!—may seem too good to be true. Yet pursuing the question *Why is there something rather than nothing?* is bound to leave our feelings about the world and our own place within it transformed. The astonishment we feel at its sheer existence may evolve into a new kind of awe as we begin to descry, if only in the faintest outlines, the reason behind that existence. Our mild anxiety about the precariousness of being may give way to confidence in a world that turns out to be coherent, luminous, and intellectually secure. Or it might yield to cosmic terror when we realize that the whole show is a mere ontological soap bubble that could pop into nothingness at any moment, without the slightest warning. And our present sense of the potential reach of human thought may give way to a newfound humility at its limits, or to a newfound wonder at its leaps and bounds—or a bit of both. We may feel like the mathematician Georg Cantor did when he made a profound new discovery about infinity. "I see it," Cantor exclaimed, "but I don't believe it."

Before we start delving into the mystery of existence, it seems only fair to give nothingness its due. For, as the German diplomat and philosopher Max Scheler wrote, "He who has not, as it were, looked into the abyss of the absolute Nothing will completely overlook the eminently positive content of the realization that there is something rather than nothing."

Let us, then, dip briefly into that abyss, with full assurance that we will not come up empty-handed. For, as the old saying goes: Nothing seek, nothing find.

Interlude

The Arithmetic of Nothingness

Mathematics has a name for nothing, and that is "zero." It is notable that the root of zero is a Hindu word: *sunya*, meaning "void" or "emptiness." For it was among Hindu mathematicians that our notion of zero arose.

To the Greeks and Romans, the very idea of zero was inconceivable—how could a nothing be a something? Lacking a symbol for it in their number systems, they could not take advantage of convenient "positional" notation (in which, for example, 307 stands for 3 hundreds, no tens, and 7 ones). That's one reason why multiplying with roman numerals is hell.

The idea of emptiness was familiar to Indian mathematicians from Buddhist philosophy. They had no difficulty with an abstract symbol that signified nothing. Their notation was transmitted westward to Europe during the Middle Ages by Arab scholars—hence our "arabic numerals." The Hindu *sunya* became the Arabic *sifr*, which shows up in English in both the words "zero" and "cipher."

Although European mathematicians welcomed zero as a notational device, they were at first chary of the concept behind it. Zero was initially regarded more as a punctuation mark than as a number in its own right. But it soon began to take on greater reality. Oddly enough, the rise of commerce had something to do with this. When double-entry bookkeeping was invented in Italy around 1340, zero came to be viewed as a natural dividing point between credits and debits.

Whether discovered or invented, zero was clearly a number to be reckoned with. Philosophical doubts about its nature receded before the virtuoso calculations of mathematicians such as Fibonacci and Fermat. Zero was a gift to algebraists when it came to solving equations: if the equation could be put in the form $ab = 0$, then one could deduce that either $a = 0$ or $b = 0$.

As for the origin of the numeral "0," that has eluded historians of antiquity. On one theory, now discredited by scholars, the numeral comes from the first letter of the Greek word for "nothing," *ouden*. On another theory, admittedly fanciful, its form derives from the circular impression left by a counting chip in the sand—the presence of an absence.

Suppose we let 0 stand for Nothing and 1 stand for Something. Then we get a sort of toy version of the mystery of existence: How can you get from 0 to 1?

In higher mathematics, there is a simple sense in which the transition from 0 to 1 is impossible. Mathematicians say that a number is "regular" if it can't be reached via the numerical resources lying below it. More precisely, the number n is regular if it cannot be reached by adding up fewer than n numbers that are themselves smaller than n.

It is easy to see that 1 is a regular number. It cannot be reached from below, where all there is to work with is 0. The sum of zero 0's is 0, and that's that. So you can't get from Nothing to Something.

Curiously, 1 is not the only number that is unreachable in this way. The number 2 also turns out to be regular, since it can't be reached by adding up fewer than two numbers that are less than 2. (Try it and see.) So you can't get from Unity to Plurality.

The rest of the finite numbers lack this interesting property of regularity. They *can* be reached from below. (The number 3, for example, can be reached by adding up two numbers, 1 and 2, each of which is itself less than 3.) But the first infinite number, denoted by the Greek letter omega, does turn out to be regular. It can't be reached by summing up any finite collection of finite numbers. So you can't get from Finite to Infinite.

But back to 0 and 1. Is there some other way of bridging the gap between them—the arithmetical gap between Nothing and Something?

As it happens, no less a genius than Leibniz thought he had found a bridge. Besides being a towering figure in the history of philosophy, Leibniz was also a great mathematician. He invented the calculus, more or less simultaneously with Newton. (The two men feuded bitterly over who was the true originator, but one thing is certain: Leibniz's notation was a hell of a lot better than Newton's.)

Among much else, the calculus deals with infinite series. One such infinite series that Leibniz derived is:

$$1/(1-x) = 1 + x + x^2 + x^3 + x^4 + x^5 + \ldots$$

Showing remarkable sangfroid, Leibniz plugged the number –1 into his series, which yielded:

$$1/2 = 1-1 + 1-1 + 1-1 + \ldots$$

With appropriate bracketing, this yielded the interesting equation:

$$1/2 = (1-1) + (1-1) + (1-1) + \ldots$$

or:

$$1/2 = 0 + 0 + 0 + \ldots$$

Leibniz was transfixed. Here was a mathematical analogue of the mystery of creation! The equation seemed to prove that Something could indeed issue from Nothing.

Alas, he was deceived. As mathematicians soon came to appreciate, such series made no sense unless they were *convergent* series—unless, that is, the infinite sum in question eventually homed in on a single value. Leibniz's oscillating series failed to meet this criterion, since its partial sums kept jumping from 0 to 1 and back again. Thus his "proof" was invalid.

The mathematician in him must surely have suspected this, even as the metaphysician in him rejoiced.

But perhaps something can be salvaged from this conceptual wreckage. Consider a simpler equation:

$$0 = 1-1$$

What might it represent? That 1 and –1 add up to zero, of course.

But that is interesting. Picture the reverse of the process: not 1 and –1 coming together to make 0, but 0 peeling apart, as it were, into 1 and –1. Where once you had Nothing, now you have *two* Somethings! Opposites of some kind, evidently. Positive and negative energy. Matter and antimatter. Yin and yang.

Even more suggestively, –1 might be thought of as the same entity as 1, only moving *backward in time*. This is the interpretation seized on by the Oxford chemist (and outspoken atheist) Peter Atkins. "Opposites," he writes, "are distinguished by their direction of travel in time." In the absence of time, –1 and 1 cancel; they coalesce into zero. Time allows the two opposites to peel apart—and it is this peeling apart that, in turn, marks the emergence of time. It was thus, Atkins proposes, that the spontaneous creation of the universe got under way. (John Updike was so struck by this scenario that he used it in the conclusion of his novel *Roger's Version* as an alternative to theism as an explanation for existence.)

All that from 0 = 1–1. The equation is more ontologically fraught than one might have guessed.

Simple arithmetic is not the only way that mathematics can build a bridge between Nothingness and Being. Set theory also furnishes the materials. Quite early in their mathematical education, indeed often in grade school, children are introduced to a curious thing called the "empty set." This is a set that has no members at all—like the set of female U.S. presidents preceding Barack Obama. It is conventionally denoted by {}, the set brackets with nothing inside of them, or by the symbol Ø.

Children sometimes bridle at the idea of the empty set. How, they ask, can a collection that contains nothing really be a collection? They are not

alone in their skepticism. One of the greatest mathematicians of the nineteenth century, Richard Dedekind, refused to regard the empty set as anything more than a convenient fiction. Ernst Zermelo, a creator of set theory,
called it "improper." More recently, the great American philosopher David
K. Lewis mocked the empty set as "a little speck of sheer nothingness, a
sort of black hole in the fabric of Reality itself . . . a special individual with
a whiff of nothingness about it."

Does the empty set exist? Can there be a Something whose essence—
indeed, whose *only* feature—is that it encompasses Nothing? Neither
believers nor skeptics have produced any strong arguments for or against
the empty set. In mathematics it is simply taken for granted. (Its existence
can be proved from the axioms of set theory, on the assumption that there
is at least one other set in the universe.)

Let's be metaphysically liberal and say that the empty set does exist.
Even if there's nothing, there must be a set that contains it.

Admit that, and a regular ontological orgy gets under way. For, if the
empty set Ø exists, so does a set that contains it: {Ø}. And so does a set that
contains both Ø and {Ø}: {Ø, {Ø}}. And so does a set that contains that new
set, plus Ø and {Ø}: {Ø, {Ø}, {Ø, {Ø}}}. And on and on.

Out of sheer nothingness, a remarkable profusion of entities has come
into being. These entities are not made out of any "stuff." They are pure,
abstract structure. They can mimic the structure of the numbers. (In the
preceding paragraph, we "constructed" the numbers 1, 2, and 3 out of the
empty set.) And numbers, with their rich web of interrelations, can mimic
complicated worlds. Indeed, they can mimic the entire universe. At least
they can if, as thinkers like the physicist John Archibald Wheeler have
speculated, the universe consists of mathematically structured information. (This view is captured by the slogan "it from bit.") The whole show of
reality can be generated out of the empty set—out of Nothing.

But that, of course, presumes there is Nothing to start with.

3

A BRIEF HISTORY OF NOTHING

Hartley told his Mother, that he was thinking all day—all the morning, all the day, all the evening—"what it would be, if there were
Nothing! if all the men, & women, & Trees, & grass, and birds &
beasts, & the Sky, & the Ground, were all gone: *Darkness & Cold-
ness*—& nothing to be dark & cold."
—SAMUEL TAYLOR COLERIDGE, *letter to Sara ("Asra") Hutchinson,
June 1802 (Hartley was Coleridge's son.)*

NOTHING! thou elder brother even to shade
That hadst a being ere the world was made,
And (well fixed) are alone of ending not afraid.
—JOHN WILMOT, EARL OF ROCHESTER, "U*pon Nothing*"

Nothing,
said Heidegger,
the modernist
eminence,
noths.
—ARCHILOCHUS JONES, *"Metaphysics Explained for You"*

What is nothing? Macbeth answered this question with admi-
rable concinnity: "Nothing is, but what is not." My dictionary puts it somewhat more paradoxically—"*nothing* (n.): a
thing that does not exist." Although Parmenides, the ancient Eleatic sage,
declared that it was impossible to speak of what is not—thereby violating

his own precept—the plain man knows better. Nothing is popularly held to be better than a dry martini, but worse than sand in the bedsheets. A poor man has it, a rich man needs it, and if you eat it for a long time, it'll kill you. On occasion, nothing could be further from the truth, but it is not clear how much further. It can be both black and white all over at the same time. Nothing is impossible for God, yet it is a cinch for the rankest incompetent. No matter what pair of contradictory properties you choose, nothing seems capable of embodying them. From this it might be concluded that nothing is mysterious. But that would only mean that everything is obvious—including, presumably, nothing.

That, perhaps, is why the world abounds with people who know, understand, and believe in nothing. But beware of speaking blasphemously of nothing, for there are also many bumptious types about—call them "nullophiles"—who are fond of declaring that, to them, nothing is sacred.

Ex nihilo nihil fit, averred the ancient philosophers, and King Lear agreed: nothing comes of nothing. This maxim would appear to attribute to nothing a remarkable power: that of generating itself—of being, like God, *causa sui*. The philosopher Leibniz paid nothing another compliment when he observed that it was "simpler and easier than something." (Hard experience teaches the same lesson: nothing is simple, nothing is easy.) Indeed, it was the alleged simplicity of nothing that moved Leibniz to ask why there is something rather than nothing. If there *were* nothing, after all, there would be nothing to be explained—and no one to demand an explanation.

If nothing is so simple, so natural, then why, one wonders, does it seem so deeply mysterious? In the 1620s, John Donne, speaking from the pulpit, furnished a plausible answer: "The less anything is, the less we know it: how invisible, how unintelligible a thing, then, is this *Nothing*!"

And why should such a simple (albeit unintelligible) thing strike others as so sinister? Take the Swiss theologian Karl Barth, one of the most profound and brave thinkers of the twentieth century. What, asked Barth, is Nothing? It is "that which God does not will." In his massive and unfinished life's work, *Church Dogmatics*, Barth wrote, "The character of nothingness derives from its ontic peculiarity. It is evil." Nothing rose up simultaneously with Something when God created the world, according to

Barth. The two are rather like a pair of ontological twins, though contrary in moral character. It is nothingness that accounts for man's perverse tendency to do evil, to rebel against divine goodness. For Barth, nothingness was downright Satanic.

The existentialists, though godless themselves, regarded nothingness with similar dread. "Nothingness haunts being," declared Jean-Paul Sartre, in his ponderous treatise *Being and Nothingness*. For Sartre, the world was like a little sealed container of being floating on a vast sea of nothingness. Not even a Parisian café—on a good day a "fullness of being," with its booths and mirrors, its smoky atmosphere, animated voices, clinking wine glasses, and rattling saucers—could afford sure refuge from nullity. Sartre drops into the Café de Flore to keep a rendezvous with his friend Pierre. But Pierre is not there! *Et voilà*: a little pool of nothingness has seeped into the realm of being from the great *néant* that surrounds it. Since it is through dashed hopes and thwarted expectations that nothingness intrudes into the world, our very consciousness must be to blame. Consciousness, says Sartre, is nothing less (or more?) than a "hole at the heart of being."

Sartre's fellow existentialist Martin Heidegger was filled with *Angst* at the very thought of nothing, although this did not keep him from writing copiously about it. "*Anxiety reveals the Nothing,*" he observed—his italics. Heidegger distinguished between fear, which has a definite object, and anxiety, a vague sense of not being at home in the world. What, in our anxious states, are we afraid of? Nothing! Our existence issues from the abyss of nothingness and ends in the nothingness of death. Thus the intellectual encounter each of us has with nothingness is suffused with the dread of our own impending nonbeing.

As to the nature of nothingness, Heidegger was wildly vague. "Nothing is neither an object nor anything that is at all," he sensibly declared at one point. Yet in order to avoid saying, *Das Nichts ist*—"Nothing is"—he was driven to an even more peculiar locution, *Das Nichts nichtet*: "Nothing noths." Instead of being an inert object, nothingness would appear to be a dynamic thing, a sort of annihilating force.

The American philosopher Robert Nozick took Heidegger's idea a step further. If nothing is an annihilating force, Nozick conjectured, it might just "noth" itself, thereby giving rise to a world of being. He imagined nothing as "a vacuum force, sucking things into non-existence or keeping them there. If this force acts upon itself, it sucks nothingness into nothingness, producing something or, perhaps, everything." Nozick recalled the vacuum cleaner–like beast in the movie *Yellow Submarine* that goes around sucking up all it encounters. After hoovering away everything else on the movie screen, it ultimately turns on itself and sucks itself into nonexistence. With a pop, the world reappears, along with the Beatles.

Nozick's speculations about nothing, though playful in spirit, left some of his fellow philosophers exasperated. They felt he was willfully sliding into nonsense. One of them, the Oxford philosopher Myles Burnyeat, commented, "By the time one has struggled through this wild and woolly attempt to find a category beyond existence and non-existence, and marvelled at such things as the graph showing 'the amount of Nothingness Force it takes to nothing some more of the Nothingness Force being exerted,' one is ready to turn logical positivist on the spot."

The logical positivists, indeed, dismissed all such speculation as much ado about nothing. One of the most distinguished of them, Rudolf Carnap, observed that the existentialists had been fooled by the grammar of "nothing": since it behaves like a noun, they assumed it must refer to an entity—a something. This is the same blunder that the Red King makes in Lewis Carroll's *Through the Looking-Glass*: if Nobody had passed the messenger on the road, the Red King reasoned, then Nobody must have arrived first. Treating "nothing" as the name of a thing allows one to generate endless paradoxical twaddle, as the opening paragraphs of this very chapter attest.

THE IDEA THAT it is nonsensical to talk about nothing goes back to the dawn of Western philosophy. It was Parmenides, the greatest of the pre-Socratics, who was most emphatic on this point. Parmenides is a somewhat mysterious figure. A native of Elea, in southern Italy, he flourished in the mid-fifth century BCE. As an elderly man, he reputedly met the young Socrates. Plato described him as "venerable and awesome." Parmenides was the first Greek philosopher to set out a sustained logical argument about the nature of reality, and thus he might be regarded as the original metaphysician. Curiously, he chose to present his argument in the form of a long allegorical poem, of which some 150 lines survive. In the poem, an unnamed goddess offers the narrator a choice between two paths: the path of being, and the path of nonbeing. But the latter path proves to be illusory, since nonbeing can be neither thought about nor spoken of. Just as "seeing nothing" is not seeing, speaking or thinking of nothing is not speaking or thinking at all, and approaching nothing is failing to make progress.

The Parmenidean line certainly seems to deflate the mystery of existence. If we cannot speak meaningfully of "nothing," then we cannot meaningfully ask why there is something rather than nothing. The words would have no more sense than the bubbles issuing from the mouth of a fish.

But sense can be quickly restored by drawing a simple distinction between *nothing* and *nothingness*. As the logicians remind us, *nothing* is not a name; it is mere shorthand for "not anything." To say, for example, that "nothing is greater than God" is not to talk about a super-divine entity; it is simply to say that there is not anything greater than God. "Nothingness," by contrast, is indeed a name. It designates an ontological option, a possible reality, a conceivable state of affairs: that in which nothing exists.

Some languages mark the distinction between *nothing* and *nothingness* more clearly than others. In French, for example, "nothing" is *rien*, while "nothingness" is *le néant*. In mathematics, the distinction is made precise by the notion of the "empty set." An empty set is a set that has no members; hence it is a something that contains nothing. Using the brackets of set theory, one gets the following equations:

$$Le \ néant = \{rien\}$$

$$Nothingness = \{nothing\}$$

Once *nothing* and *nothingness* are distinguished, it is easy to resolve the supposed paradoxes about nothing that arise from conflating the two, like those the ancient Greek philosophers were so fond of. ("How can anything be something that is not something?" one Greek riddle went. "By being nothing.") It is also easy to deal with gnomic formulations like Heidegger's *Das Nichts nichtet*. If Englished as "Nothing noths," the statement is quite true but uninteresting: of course there is not anything that "noths"! If Englished as "Nothingness noths," then it is quite false. Nothingness does nothing of the kind. It is merely a possible reality, and a possible reality either can be the case or fail to be the case. That is all. It cannot engage in any activity; it can neither cause nor "noth."

But *is* nothingness a possible reality? Certainly we have all experienced absence and loss. We are intimately familiar with holes and gaps, with lacks and deficits. Indeed, as the late Peter Heath, a mischievous British philosopher (and former teacher of mine), observed, voids and vacancies are even advertised in the newspapers. But these are mere bits of nothingness, surrounded, as they are, by a world of being. What about Absolute Nothingness, the total absence of *everything*? Is this possible?

Some philosophers have argued that it isn't. The very idea, they say, is self-contradictory. If these philosophers are right, then the riddle of being has a cheap and rather trivial solution: there is something rather than nothing simply because *nothingness is impossible*. As one contemporary philosopher has put it, "There is just no alternative to being."

Could that be true? Close your eyes, if you will, and stop up your ears. Now picture to yourself an absolute void. Try to wish into nonbeing the entire contents of the world. You might begin, as Coleridge's little boy did, by imagining away all the men and women and trees and grass and birds and beasts and earth and sky. And not just the sky, but everything in it. Think of the lights going out all over the cosmos: the sun disappearing, the stars extinguished, the galaxies winking into nonexistence one by one, or billion by billion. In your mind's eye, the entire cosmos is sliding into silence, cold, and darkness—with nothing to be silent or cold or dark. You have succeeded in imagining absolute nothingness.

Or have you? When the French philosopher Henri Bergson tried to imagine universal annihilation, he found that there was inevitably something left over at the end of the experiment: his inner self. Bergson thought of the world of being as "an embroidery on the canvas of the void." But when he attempted to strip this embroidery away, the canvas of his consciousness remained. Try as he might, he could not suppress it. "At the very instant that my consciousness is extinguished," he wrote, "another consciousness lights up—or rather, it was already alight; it had arisen the instant before, in order to witness the extinction of the first." He found it impossible to imagine absolute nothingness without some residuum of consciousness creeping into the darkness, like a little light under the door. Therefore, he concluded, nothingness must be an impossibility.

Bergson was not the only philosopher to argue in this way. The British idealist F. H. Bradley, author of the dauntingly titled *Appearance and Reality*, similarly maintained that sheer nothingness was unthinkable. He too concluded that it must therefore be impossible.

One of the more confused attempts to imagine nothingness was made by "S," a patient of the distinguished Russian psychologist Aleksandr Luria. S had such an extraordinary memory that Luria wrote an entire book about him, titled *The Mind of a Mnemonist*. Oddly though, his memory was almost purely visual. So when S tried to conceive of nothingness, the experiment went disastrously awry:

> In order for me to grasp the meaning of a thing, I have to see it.
> ... Take the word *nothing* ... I see this *nothing* and it is something.
> ... So I turned to my wife and asked her what *nothing* meant. ...
> She simply said: "*Nothing* means there is nothing." I understood it
> differently. I saw this *nothing*. ... If *nothing* can appear to a person,
> that means it is something. That's where the trouble comes in.

Perhaps any attempt to summon up an image of nothing is self-defeating. Even so, is *thinkability* a reliable test for *possibility*? Does the fact that we cannot imagine absolute nothingness—except, perhaps, in a state of dreamless sleep—mean that something or other must perforce exist?

One must beware here of falling into what has been called the *philosopher's fallacy*: a tendency to mistake a failure of the imagination for an insight into the way reality has to be. "I can't imagine it otherwise," a thinker prone to this fallacy says to himself; "therefore it must be so." There are many things that lie beyond the powers of our imagination that are not only possible but real. We can't visualize colorless objects, for example, yet atoms are colorless. (They are not even gray.) Most of us, with the exception of a few preternaturally gifted mathematicians, cannot imagine curved space. Yet Einstein's relativity theory tells us that we actually live in a curved four-dimensional spacetime manifold, one that violates Euclidean geometry—something that Immanuel Kant found unimaginable and thus ruled out on philosophical grounds.

Bergson and Bradley thought that absolute nothingness was self-contradictory, because the very possibility would entail the existence of an observer to think about it. Call this the "observer argument" against nothingness. The observer argument is not only dubious on general grounds, but also has some wild implications. It means that *every* possible world must contain at least one conscious observer. But surely a universe without consciousness is physically possible. If the constants of nature in our own universe—the strength of the weak nuclear force, the mass of the top quark, and so on—were even slightly different from their actual values, there would have been no evolution of life in the universe, just a lot of brute matter. But, by the logic of the observer argument, such a zombie universe would be impossible, since there would be no one to observe it.

Bergson's version of the observer argument has a still more absurd implication. In his mind's eye, he could not abolish his own self. On the principle that what is unimaginable is impossible, he ought to have concluded that his *own* nonexistence was impossible: that no matter how reality had turned out—empty, full, whatever—it was metaphysically guaranteed to include Monsieur Bergson; that he himself was a God-like necessary being. To call this solipsism would be charitable.

There is a second argument against nothingness that, though similar in its logic, runs along more objective lines. Like the observer argument, it too says that our effort to imagine absolute nothingness is doomed to be only partial, never complete. But instead of pointing to consciousness as the thing left over, it cites a residuum that is nonmental. Even when all the contents of the cosmos have been imaginatively banished, the argument goes, we are always left with the abstract setting that they inhabited. This setting may be empty, but it is not nothing. A container with no contents is still a container. Let's call this the "container argument" against nothingness.

One venerable exponent of the container argument is Bede Rundle, a contemporary philosopher at Oxford. "Our attempt to think away every-thing amounts to envisaging a region of space which has been evacuated of its every occupant, an exercise which gives no more substance to the pos-sibility of there being nothing than does envisaging an empty cupboard," Rundle has written (in a book tellingly titled *Why There Is Something Rather Than Nothing*). And just what is this "empty cupboard"? Rundle appears to be identifying it with space itself. Since one cannot "think away" the presence of space, he suggests, it must be part of any possible reality—a necessary existent, like God, or Henri Bergson's inner self.

So is space to be our great bulwark against nothingness? Rundle hedges his bets. At one point he considers an alternative argument, to the effect that the very idea of nothingness is incoherent. If there were nothing, then it would have been a *fact* that there was nothing. So at least one thing would exist after all: that fact! (This is a truly terrible argument; the enumera-tion of its fallacies is left as an exercise for the reader.) But it is space that Rundle keeps coming back to, since he just can't think it away, try as he might. "Space is not *nothing*," he insists, "it is something you can stare into or travel through, something of which there can be volumes."

Not everyone shares Rundle's conviction that space is a something. Among philosophers, there are two competing views of what space actu-ally is. (To be scientifically up-to-date, we should be talking about "space-time" rather than "space," but no matter.) One of them, the *substantival* view, goes back to Newton. It holds that space is indeed a real thing, with its own intrinsic geometry, and that it would continue to exist even if all its contents vanished. The other view of space, the *relational* view, goes back to Newton's great rival, Leibniz. It holds that space is not a thing unto itself, but merely a web of relations among things. Space could no more exist apart from the things that it relates, on Leibniz's view, than the grin of the Cheshire Cat could exist apart from the feline itself.

The ontological debate between the Newtonians and the Leibnizians continues to the present day, and it's a lively one. Relativity theory, in which spacetime affects the behavior of matter, has tipped the balance somewhat in favor of the substantivalists.

But it's not necessary to resolve this debate to see whether the container argument is any good. Suppose the relationists are right, and space is just a convenient theoretical fiction. In that case, if the contents of the cosmos were to vanish, space would vanish along with them, leaving absolute nothingness.

Now suppose, contrariwise, that the substantivalists are right. Suppose that space is a genuine cosmic arena, with an existence all its own. Then this arena could survive the disappearance of its material contents. Even with everything gone, there would still be unoccupied positions. But if space has real objective existence, so does its geometrical form. It could be infinite in extent. But it could also be finite, even though it has no boundary. (The surface of a basketball, for example, is a finite two-dimensional space that has no boundary.) Such "closed" spacetimes are consistent with Einstein's relativity theory. Indeed, Stephen Hawking and other cosmologists have theorized that the spacetime of our own universe is finite and unbounded, like a higher-dimensional analogue to the surface of a basketball. In that case, it is not hard to "think away" spacetime along with everything in it. Just imagine that basketball deflating, or rather shrinking. In your mind's eye, the finite radius of the basketball-cosmos grows smaller and smaller until, finally, it reaches zero. Now the spacetime arena itself has vanished, leaving absolute nothingness behind.

This thought experiment leads to an elegant scientific definition (originally due to the physicist Alex Vilenkin):

Nothingness = a closed spherical spacetime of zero radius

So the container argument fails, regardless of what the nature of the container might turn out to be. If spacetime is not a genuine entity, but merely a set of relations among things, then it vanishes along with those things and hence is no obstacle to the possibility of nothingness. If spacetime *is* a genuine entity, with its own peculiar structure and quiddity, then it can be "disappeared" by the imagination just like the rest of the furniture of reality.

Voiding reality in the mind's eye is a purely imaginative achievement. What if one tried to carry it out in the lab? Aristotle thought that this would be impossible. He produced a variety of arguments, both empirical and conceptual, purporting to show that you can't empty out a region of space. The Aristotelian orthodoxy that "nature abhors a vacuum" held until the middle of the seventeenth century, when it was decisively overthrown by one of Galileo's pupils, Evangelista Torricelli. An ingenious experimentalist, Torricelli had the happy idea of pouring mercury into a test tube, and then, with his finger over the open end, plunging it into a mercury bath. With the tube standing vertically upside down, a little airless void appeared above the column of mercury. What Torricelli had done was to create the first barometer. He had also demonstrated that nature's supposed *horror vacui* was really nothing more than the weight of the atmospheric air pressing down on us.

But did Torricelli succeed in producing a bit of true nothingness? Not quite. Today, we know that the sort of airless space he was the first to create is far from being completely empty. The most perfect vacuum, it turns out, still contains something. In physics, the notion of "something" is quantified by energy. (Even matter, as Einstein's most famous equation shows, is just frozen energy.) Physically speaking, space is as empty as it can be when it is devoid of energy.

Now, suppose you try to remove every bit of energy from a region of space. Suppose, in other words, you try to reduce that region to its state of lowest energy, which is known as its "vacuum state." At a certain point in this energy-draining process, something very counterintuitive will occur. An entity that physicists call the "Higgs field" will spontaneously emerge. And this Higgs field cannot be got rid of, because its contribution to the total energy of the space you are trying to empty out is actually *negative*. The Higgs field is a "something" that contains less energy than a "nothing." And it is accompanied by a riot of "virtual particles" that ceaselessly wink in and out of existence. Space in a vacuum state turns out to be very busy indeed, rather like Times Square on New Year's Eve.

PHILOSOPHERS WHO BELIEVE in Nothing—they sometimes call themselves "metaphysical nihilists"—try to steer clear of such physical snags. In the late 1990s, several British and American philosophers jointly pioneered what has become known as the "subtraction argument." Unlike the observer and container arguments, which were anti-nothingness, the subtraction argument is pro-nothingness. It is meant to demonstrate that an absolute void is a genuine metaphysical possibility.

The subtraction argument begins by assuming, plausibly enough, that the world contains a *finite* number of objects—people, tables, chairs, rocks, and so forth. It also assumes that each of these objects is *contingent*: although the object does in fact exist, it might not have existed. This too seems plausible. Think of the movie *It's a Wonderful Life*, and its protagonist George Bailey (played by Jimmy Stewart). After a series of setbacks in his life, George finds himself contemplating suicide. Thanks to the intervention of an angel named Clarence, George gets to see what the world would have looked like if he had never been born. He is confronted with the contingency of his own existence. The same contingency seems to infect not only individual people, but also the entire inventory of actually existing things, from the Milky Way to the Eiffel Tower to the dog sleeping on your sofa to the speck of dust on the mousepad of your laptop. Each of these things, although it happens to exist, might not have existed if the cosmos had unfolded differently. Finally, the subtraction argument makes an assumption of *independence*: that the nonexistence of one thing does not necessitate the existence of anything else.

With all three of these premises in place—finiteness, contingency, and independence—it is easy to derive the conclusion that there might have been nothing at all. You simply subtract each contingent object from the world, one by one, until you end up with absolute emptiness, a pure void. This "subtraction" is supposed to be metaphorical rather than literal. Each stage of the argument asserts a relationship between possible worlds: if a world with n objects is possible, then a world with $n-1$ objects is also possible. At the penultimate stage in the subtraction process, the world might consist of no more than a single grain of sand. If such a sad little world is possible, then so is a world where that grain of sand is deleted—a world of nothingness.

The subtraction argument is generally considered to be the strongest one in the arsenal of the metaphysical nihilists. Indeed, it may be the only positive argument they have. Although I have stated the argument somewhat crudely, its proponents have painstakingly put it into a form where it appears to be logically valid: no mean feat. If the premises are true, the conclusion—that absolute nothingness is possible—must also be true.

But *are* the premises of the subtraction argument true? In other words, is the argument not merely valid, but also (as logicians say) *sound*?

Well, the finiteness and contingency premises seem okay. But the third premise, that of independence, is more dubious. Can we really be sure that the nonexistence of one thing does not entail the existence of something else? Think again of *It's a Wonderful Life*. In the alternative world where George Bailey never existed, many other possible things did exist as a consequence—like the sleazy bars and pawnshops of "Pottersville," which the greedy banker Mr. Potter would have created had decent George not been around to stop him. Contingent things are not so independent after all. Each thing, no matter how shaky its own claim to existence, seems implicated in a web of ontic interdependency with many other things, both actual and possible.

If a cinematic example is too fanciful for you, consider a more austere, scientific one. Suppose the world consisted of just two objects, an electron and a positron in mutual orbit. Relative to this "pair world," is there a possible "singleton world" in which only the positron exists? One might think so. But the move from the pair world to the singleton world would violate one of the bedrock principles of physics: the law of charge conservation. The net charge of the pair world sums up to zero, since the positron has charge +1 and the electron has charge −1. The net charge of the singleton world is +1. So moving from the pair world to the singleton world is tantamount to the creation of a net charge—a physical impossibility. Although the electron and positron are individually contingent, each is existentially linked to the other by the law of charge conservation.

Then how about moving from the pair world directly to nothingness? Alas, that's not physically possible either, for eliminating the electron-positron pair would violate another bedrock principle of physics: the law of mass-energy conservation. Some new entity—a photon, say, or another particle-antiparticle pair—would have to appear in their wake, as a matter of physical necessity.

The hitch here seems to be the same one that both Bergson and Rundle encountered, but in a different guise. In all three cases, absolute nothingness is thought of as a *limit*, one to be approached from the world of being. Bergson tried to approach it by imaginatively annihilating the contents of the world, only to be left with his own consciousness. Rundle tried a similar imaginative route and also fell short of the goal, ending up with an empty spatial container. Both concluded that absolute nothingness was inconceivable. The subtraction argument tries a different tack, seeking to reach nothingness via a series of logical moves. But the reasonable-sounding intuition behind the subtraction argument—*if there are some objects, there could have been fewer of them*—runs afoul of a set of fundamental physical principles: the laws of conservation. And even if those laws were somehow suspended, it is by no means clear that the world's ontological census could be steadily reduced by decrements of one, all the way down to zero. Perhaps the absence of one thing, in either imagination or reality, always entails the presence of another. Delete George Bailey from the scheme of things and up pops Pottersville.

The apparent moral is this: it's no simple matter to get from Something to Nothing. The approach is asymptotic at best, always falling short of the limit, always leaving some remainder of being, however infinitesimal. But is that surprising? To succeed in reaching Nothing from Something, after all, would be to solve the riddle of being in reverse. Any logical bridge from one to the other would presumably permit two-way traffic.

If it seems easier in the imagination to get from Something to Nothing than the reverse, that's because both the starting point and the terminus are known in advance. Suppose you sit down at a computer terminal in the reading room of the New York Public Library on Forty-second Street. There's a single character on the screen—say, "$." You press the *delete* button, and the screen becomes blank. You have effected a transition from Something to Nothing. Now suppose you happen to sit down at a terminal with a blank screen. How do you go from Nothing to Something? By pressing the *undelete* button. When you do so, however, you have no idea what will appear on the screen. Depending on what the previous user was up to, it could be a lapidary message or a mere jumble of characters. The transition from Nothing to Something seems mysterious, because you never know what you're going to get. And the same is true at the cosmic level. The Big Bang—the physical transition from nothing to something—was not only inconceivably violent, but also inherently lawless. Physics tells us that there is in principle no way of predicting what might pop out of a naked singularity. Not even God could know.

Instead of struggling to cross an impassable conceptual divide between Something and Nothing, it might be more profitable to forget about the world of being and concentrate on nothingness instead. Can absolute nothingness be coherently described, without falling into some kind of contradiction? If it can be, this might raise our confidence that it's a genuine metaphysical possibility.

But defining absolute nothingness can be a tricky business. As a first shot, one might start with this proposition:

Nothing exists.

Translated into formal logic, this becomes:

For every *x*, it is not the case that *x* exists.

Already we have a problem: "exists" does not name a property, one that things might either have or fail to have. It makes sense to say, "Some tame tigers growl and some do not." But it makes no sense to say, "Some tame tigers exist and some do not."

If we limit ourselves to proper predicates—"is blue," "is bigger than a breadbox," "is smelly," "is negatively charged," "is all-powerful," and so forth—then the task of defining absolute nothingness appears to become much messier. Now we need an unbounded, perhaps infinite list of propositions to pin down the null possibility: "There is nothing that is blue," "There is nothing that is smelly," "There is nothing that is negatively charged," and so forth. Each of these propositions has the form:

For every x, it is not the case that x is A.

Or, more concisely:

There are no As.

Each proposition in the list will rule out the existence of all objects with a certain property: all blue things, all smelly things, all negatively charged things, and so on.

If our list of nonexistents contains a proposition for every metaphysically possible property, then we will have succeeded in defining absolute nothingness by this *via negativa*. But how can we be sure that the list is exhaustive? A single omission will defeat the nullity project, by allowing the existence of some category of objects that either we forgot about or that is currently beyond our imagination. If we were drawing up the list a century ago, for example, we would have left off the proposition "For every x, it is not the case that x is a black hole."

One might try to get around this problem of exhaustiveness by dividing all possible types of things into a few fundamental categories. Descartes, for example, divided the world of being into just two kinds of substance: minds, whose essence is thought, and physical bodies, whose essence is extension. So we might try to define absolute nothingness by the pair of propositions "There are no mental things" and "There are no physical things." This neat pair would rule out the existence of consciousness, souls, angels, and deities, along with electrons, rocks, trees, and galaxies. But would it rule out the existence of mathematical entities, like numbers? Or abstract universals, like justice? Such things seem to be neither mental nor physical, yet their existence would certainly seem to spoil the state of absolute nothingness. And there may be a whole range of other possible substances, other species of being, undreamt of either by Descartes or by us.

There is one property, however, that every conceivable object, whether animal, vegetable, mineral, mental, spiritual, mathematical, or anything else, is guaranteed to possess. And that is self-identity. I have the property of being me. You have the property of being you. And so on. Indeed, "identity" is defined in logic as the relation that each and every thing bears to itself and to no other thing. In other words, it is a logical truth that:

$$\text{For every } x, x = x.$$

To exist, therefore, is to be self-identical.

With the identity relation, the statement "Something exists" becomes:

$$\text{There is an } x \text{ such that } x = x.$$

So, to capture absolute nothingness in the trap of logic, all we have to do is negate this assertion. The result is:

$$\text{It is not the case that there is an } x \text{ such that } x = x.$$

Or, equivalently:

$$\text{For every } x, \text{ it is not the case that } x = x.$$

In English, this says, "Everything fails to be self-identical." The proposition is even more lapidary when expressed in the symbols of formal logic:

$$(x) \sim (x = x).$$

(The symbol "(x)" is the universal quantifier, to be read as "for every x," and "$\sim$" is the negation operator, to be read as "it is not the case that.")

So there it is, a neat little logical glyph that says, Absolutely nothing exists. But is there a possible reality that makes it true? One prominent American philosopher, the late Milton Munitz, insisted that there is not. In his book *The Mystery of Existence*, Munitz argued that the proposition asserting that something exists—"There is an *x* such that *x* is identical to itself"—is a truth of logic. Therefore, he claimed, its negation—my neat little glyph above—is "strictly meaningless."

Munitz is correct, but in a rather trivial sense. Logicians, in order to streamline their formal systems, routinely rule out nothingness. They assume that there is always at least one individual in the universe of discourse. (This makes it easier to define truth, among its other advantages.) With this expedient, the proposition "There is an *x* such that *x* is self-identical" becomes a logical truth. But it is an artificial one. As Willard Van Orman Quine, the dean of twentieth-century American philosophy, pointed out, the stipulation of a nonempty domain is "strictly a technical convenience." It carries with it "no philosophical dogma about necessary existence." Bertrand Russell went further, regarding the conventional assumption of existence as something of a blot on logic.

To get rid of this blot, logicians who agree with Russell have invented an alternative system of logic, one that does allow for the possibility of nothingness. Such a system is called "universally free logic," because it is *free* of existence assumptions about the *universe*. In a universally free logic, the empty universe is permitted, and statements asserting the existence of something or other—statements like "There is an object which is self-identical"—cease to be logical truths.

As Quine discovered, there is a remarkably easy test for truth and falsity in the empty universe. All *existential* propositions—that is, propositions beginning "There is an *x* such that . . ."—are automatically false. On the other hand, all *universal* propositions—those beginning "For every *x* . . ."—are automatically true. Why should all universal propositions be true in an empty universe? Well, take the proposition "For every *x*, *x* is red." In a world of no objects, there are certainly no objects that fail to be red. Hence, there are no counterexamples to the claim that everything *is* red. Such universal propositions are thus said to be *vacuously* true. Quine's truth test for the empty universe is a wonderful thing—or, as he preferred to put it, "a triumph of triviality." It can decide the truth of any proposition, even very complex ones. (If the proposition has both existential and universal components connected by "and" or "or," you simply apply the method of truth tables, originally invented by Wittgenstein and now familiar to every student of elementary logic.) It settles, in a consistent way, what would be true and false in an empty universe—that is, in a state of absolute nothingness. It shows that no contradiction can be derived from the assumption that nothing exists. And this is very interesting to the metaphysical nihilist. It means that absolute nothingness is self-consistent! Contrary to what so many skeptical philosophers have believed, it is a genuine logical possibility. We may not be able fully to conceive it in our imagination, but that does not mean there is anything paradoxical about it. It may sound preposterous, but it is not absurd. Logically speaking, *there might have been nothing at all*.

Let's call this possible reality the Null World, making sure to remember that it's a "world" only by ontological courtesy, as it were. Unlike other possible worlds, it involves no spacetime, no container or stage or arena of any sort. When we talk about "it," we're not taking about any kind of object; we're merely talking about one of the various ways reality might have turned out—a way neatly captured by the formula

$$(x) \sim (x = x).$$

And this formula is not itself part of the Null World—absolutely nothingness would forbid that! It is merely our way of referring to the Null World, of logically encoding what it means for nothing whatever to exist.

Logical consistency is a great virtue. But it's not the only virtue possessed by the Null World. Nothingness is also, as Leibniz was the first to point out, the *simplest* of all possible realities. Simplicity is greatly prized in science. When rival scientific theories are equally supported by the evidence, it is the simplest of them—the one that postulates the fewest causally independent entities and properties, the one least susceptible to a trimming by Occam's razor—that scientists favor. And this is not just because simpler theories are prettier, or easier to use. Simplicity is held to be a marker of intrinsic probability, of truth. It is complex realities that are thought to stand in need of explanation, not simple ones. And no possible reality is simpler than the Null World.

The Null World is also the *least arbitrary* one. Having no objects at all, its census is a nice round zero. Any alternative world will have a nonzero census. It may contain a finite number of individuals, or it may contain an infinite number. Now, unless you are a numerologist, any finite number is bound to look arbitrary. Our own universe, for instance, seems to consist of a finite population of elementary particles (the number of which is estimated to be around 10 followed by eighty zeros). In addition, there may be nonphysical individuals hanging around, like angels. If you added all these objects up, the total census of the actual world would look like a very long odometer reading on your dashboard—lots and lots of arbitrary digits. It would seem just as arbitrary if the world contained a smaller number of objects, like seventeen. Even an infinite world would be arbitrary. For there is not just one size of infinity, but many sizes—infinitely many, in fact. Mathematicians denote the different sizes of infinity by using the Hebrew letter aleph: aleph-0, aleph-1, aleph-2, and so on. If our own world turns out to have an infinite census of objects, why should it be, say, aleph-2 rather than aleph-29? Only the Null World escapes this kind of arbitrariness.

What's more, nothingness is the most *symmetrical* of realities. Many things, like faces and snowflakes, are symmetrical in a limited way. A square has lots of symmetries, because you can flip it about an axis or rotate it by ninety degrees without altering its form. A sphere has still more symmetries: any rotation at all leaves its form unchanged. Infinite space is more symmetrical yet: you can rotate it, reflect in a mirror, or shift it in any direction without changing it a bit. Our own universe is not very symmetrical on a small scale—look at what a mess your living room is! On a cosmic scale it's more symmetrical, appearing pretty much the same whatever direction you look in. But no universe, our own included, can compete with nothingness in this respect. The Null World's utter lack of particularity makes it utterly invariant under any kind of transformation. There's nothing to shift or reflect or rotate. Fearful symmetry, indeed!

But what sort of virtue is that? Well, it may be an aesthetic one. From the time of the Greeks, with their emphasis on balance and order, symmetry has been deemed a component of objective beauty. That is not to say that the Null World is the most beautiful one (although it may be to those who prefer minimalist decor or have a taste for desert landscapes). But it is the most sublime. If Being is like the blaze of the noonday sun, then nothingness is like a starless night sky, inspiring a sort of pleasurable terror in the adventurous thinker who contemplates it.

There is a final, and rather more esoteric virtue that nothingness possesses. It has to do with *entropy*. The concept of entropy is among the most fundamental in science. It explains why some changes are irreversible and why time has a direction, an "arrow" pointing from past to future. The notion of entropy arose in the nineteenth century from the study of steam engines, and originally concerned the flow of heat. Soon, however, entropy was rethought along more abstract lines, as a measure of the disorder or randomness of a system. In the twentieth century, entropy became still more abstract, merging with the idea of pure information. (When Claude Shannon was laying the foundations of information theory, he was advised by John von Neumann that if he used "entropy" in his theory he'd never lose a debate, since nobody really understands what it means.)

Everything has an entropy. The entropy of our universe, considered a closed system, is always increasing, as things move from order to disorder. That is the second law of thermodynamics. And what about Nothingness? Can it be assigned an entropy? The computation is not hard. If a system—anything from a cup of coffee to a possible world—can exist in N different states, its maximum entropy equals $\log(N)$. The Null World, being perfectly simple, has only a single state. So its maximum entropy is $\log(1)$ = 0—which also happens to equal its minimum entropy!

So Nothingness, in addition to being the simplest, the least arbitrary, and the most symmetrical of all possible realities, also has the nicest entropy profile. Its maximum entropy equals its minimum entropy equals zero. No wonder Leonardo da Vinci was moved to exclaim, perhaps somewhat paradoxically, "Among the great things which are found among us, the existence of Nothing is the greatest."

But if Nothingness is so great, why didn't it prevail over Being in the reality sweepstakes? The virtues of the Null World are manifold and unde-niable when you think about them, but they only serve to make the mystery of existence all the more mysterious.

Or so it seemed to me until, one day back in 2006, I received in the mail a wholly unexpected letter that announced, "There is no mystery of existence."

4

THE GREAT REJECTIONIST

The letter bearing the news that "there is no mystery of existence," though unexpected, did not exactly come out of the blue. A week earlier, the *New York Times* had published a review I had written of Richard Dawkins's book *The God Delusion*. In the review, I had suggested that the question *Why is there something rather than nothing?* might be the theist's final bulwark against the encroachments of science. "If there is an ultimate explanation for our contingent and perishable world," I had observed, "it would seemingly have to appeal to something that is both necessary and imperishable, which one might label 'God.'" And this observation had touched a nerve in my correspondent, a man called Adolf Grünbaum.

The name was hardly unknown to me. In the philosophical world, Adolf Grünbaum is a man of immense stature. He is arguably the greatest living philosopher of science. In the 1950s, Grünbaum became famous as the foremost thinker about the subtleties of space and time. Three decades later, he achieved a wider degree of fame—and some notoriety—by launching a sustained and powerful attack on Freudian psychoanalysis. This brought down on him the wrath of much of the psychoanalytic world and landed him on the front page of the science section of the *New York Times*.

All of this I did know about the man. What I hadn't been aware of was Grünbaum's implacable hostility to religious belief. He was particularly irked, it seemed, by cosmic mystery-mongering as a strategy for shoring up belief in a supernatural creator. As far as he was concerned, the question *Why is there a world rather than nothing at all?* was not a path to God or to anything else. It was, to borrow a term from his native German, a *Scheinproblem*—a pseudo-problem.

What made Grünbaum such a fierce rejectionist? I could understand why someone might think the mystery of existence was, by its very nature, insoluble. But to laugh it off as a pseudo-problem seemed a bit too cavalier. Still, if Grünbaum turned out to be right, the whole quest to explain the existence of the world would be a colossal waste of effort, a fool's errand. Why bother trying to solve a mystery when you can simply dissolve it? Why go on a hunt for a Snark if all that's out there is a Boojum?

So, not without trepidation, I wrote back to Grünbaum. Could we chat? He responded with characteristic brio, inviting me to come see him in Pittsburgh, where he has lived and taught for the last five decades. He'd be delighted to explain why the mystery of existence was a nonstarter, he said in his letter, even if it took a few days to convince me. When it came to his philosophical tutelage, I could "write my own ticket."

I had never been to Pittsburgh, a city I knew only from the movie *Flashdance*. But I was eager to meet Grünbaum, and to see the Monongahela River. So I caught the first flight I could from New York and, a couple of hours later, checked into a chain hotel that conveniently stood in the shadow of the University of Pittsburgh's soaring neo-Gothic Cathedral of Learning. My eager mentor Grünbaum was waiting for me in the lobby when I arrived, grinning amiably and looking like an octogenarian cross between Danny DeVito and Edward G. Robinson.

That evening, over drinks and dinner at a downtown Pittsburgh restaurant called the Common Plea, Grünbaum told me about the origins of his antipathy to theism. He traced it back to his childhood in Cologne, Germany, where he was born in 1923, during the tumultuous era of the Weimar Republic. Cologne, with its famous cathedral, was a predominantly Roman Catholic city. Grünbaum's family was part of a small Jewish minority, numbering around twelve thousand. They lived on Rubensstrasse, a street named after the Dutch painter. By the time Grünbaum was ten, the

Nazis had come to power. He vividly recalls being beaten up in the street by young thugs who announced to him that *die Juden haben unseren Heiland getötet*—"the Jews killed our Savior." He also recalls his athletic development being "psychologically stunted" because of the close association between Nazi mass rallies and athletic parades.

While still a boy, Grünbaum began to doubt the existence of God. He was repelled by the "ethically monstrous" biblical story in which Abraham is called on to sacrifice his innocent son as a test of his fealty to God. He found it absurd that there was a taboo against mentioning the name of God, Yehovah. When he blithely pronounced the word out loud in Hebrew class, the teacher pounded the table and told him it was the worst thing a Jew could do.

Grünbaum's disenchantment with religion, he told me, coincided with the beginnings of his interest in philosophy. The rabbi at the family's synagogue often alluded to Kant and Hegel in his sermons. Grünbaum was motivated to pick up an introductory book about philosophy, which, among other speculations, dealt with the origin of the universe. He also began to read Schopenhauer, admiring the philosopher both for his compassionate atheistic Buddhism and for his literary flair. By the time of Grünbaum's bar mitzvah in 1936, at the age of thirteen, he was a confirmed atheist. The next year, his family escaped Nazi Germany for the United States, fetching up in a neighborhood in southern Brooklyn. Grünbaum commuted to high school in the Bronx—an hour and a half each way on the subway—where he mastered English by means of a bilingual edition of Shakespeare's plays.

Drafted into the army during the Second World War, Grünbaum became an intelligence officer. At the age of twenty-two he was back in Germany with the American army, interrogating captured Nazis in Berlin. Among those he was in charge of questioning, I was amazed to hear, was Ludwig Bieberbach—the man behind the "Bieberbach conjecture," for decades one of the great unsolved problems in mathematics, ranking just below Fermat's last theorem. The idea that Bieberbach was an actual flesh-and-blood human—let alone one who customarily lectured to his students at the University of Berlin decked out in a Nazi SA uniform—was slightly staggering to me. Grünbaum's contempt for this Nazi mathematician was more than moral. It was also intellectual. In supporting Hitler's anti-Semitism, Bieberbach publicly argued that Nordic mathematicians took a wholesome geometrical approach to their subject, whereas the Jewish mind operated in a morbidly abstract way. The fact that Bieberbach had willfully overlooked the "glaring counterexample" to this generalization—namely, the Jewish physicist Albert Einstein, whose relativity theory showed that gravity was really geometry—enraged Grünbaum. It left him, he said, with a low threshold of indignation when it came to "sloppy, dishonest, and tendentious argument"—including arguments about why the universe exists.

Despite his advanced age and diminutive size, Grünbaum ate with a hearty appetite. He made his way through an entrée of veal and then an enormous plate of angel-hair pasta, followed by another plate of portobello mushrooms. Forgoing wine, which he said made him sick, he continued to drink Cosmopolitans ("that's my speed") through the meal, as he regaled me, in his precise diction and vestigial German accent, with philosophical gossip. When it was over, he kindly drove me back to my hotel. On the way, we passed a rather imposing church, presumably one of Pittsburgh's architectural jewels. "Do you worship there?" I asked him, trying not to sound too puckish.

"Oh, *every day*," he replied.

IN MY HOTEL room the next morning, foggily working through the formidable pile of reprinted papers from various philosophy journals the professor had given me—papers with intellectually belligerent titles like "The Poverty of Theistic Cosmology" and "The Pseudo-Problem of Creation"—I tried to fathom why Grünbaum was so disdainful of the mystery of existence. His contempt for those who took it seriously leapt off the page. They were not just "obtuse," but "exasperatingly obtuse." Their reasoning was "gross," "crude," "bizarre," and "inane," amounting to "mere farce." It was beyond "fatuous": it was "ludicrously fatuous."

It didn't take long for me to understand why he felt this way. Unlike Leibniz and Schopenhauer, unlike Wittgenstein and Heidegger and Dawkins

and Hawking and Proust, unlike any number of contemporary philosophers and scientists and theologians and just about any ordinary reflective
person, Grünbaum finds the existence of the world utterly *unastonishing*.
And he is utterly convinced that it is *rational* for him to be unastonished.

Consider again the basic mystery as originally stated by Leibniz: why is
there something rather than nothing? Grünbaum dubs this, with appropriate grandeur (and perhaps a hint of irony), the Primordial Existential
Question. But what makes it legitimate? Like any other *why* question, he
observes, it rests on hidden presuppositions. Not only does it presuppose
that there must be some explanation for the existence of the world. It also
takes for granted that the world *needs* an explanation—that, in the absence
of some overriding cause or reason, nothingness would be expected to
prevail.

But why *should* nothingness prevail? Those who profess puzzlement at
the existence of a world like ours—one teeming with life and stars and consciousness and dark matter and all kinds of stuff we haven't even discovered yet—seem to have an intellectual prejudice, one that favors the Null
World. Nothingness is the natural state of affairs, they implicitly believe,
the ontological default option. It is only deviations from nothingness that
are mysterious, that require an explanation.

And where did they get this belief in what Grünbaum derisively labels
the Spontaneity of Nothingness—a belief which seems so obvious to them
that they don't even bother to defend it? Whether they realize it or not, he
argues, they got it from religion. Even atheists like Dawkins unwittingly
imbibed it "with their mother's milk." The Spontaneity of Nothingness is a
distinctly Christian precept, Grünbaum claims. It was inspired by the doctrine of creation *ex nihilo*, which arose in the second century after Christ.
According to Christian dogma, God, being all-powerful, had no need of
any preexisting materials out of which to fashion the world. He brought it
into being out of sheer nothingness. (Presumably the Genesis account of
creation, in which God created the world by imposing order on a sort of
watery chaos, can be dismissed as mythopoetic license.)

But God, according to Christian dogma, is not only the creator of the
world. He is also its sustainer. Once created, the world is utterly dependent

on him for its continuing existence. He works around the clock to keep it
in a state of being. If God ceased existentially supporting the world, even
for a moment, it would, to use a phrase from the twentieth-century British
archbishop William Temple, "collapse into nonexistence." The world is not
like a house, which, once the builder is finished with it, continues to stand.
Rather, it is like a car balanced precariously at the edge of a cliff. Without
divine power to maintain its balance, it would plunge into the precipice
of nothingness.

The ancient Greeks did not share this Christian idea of creation *ex
nihilo*. Nor did the ancient Indian philosophers. Thus it is hardly surprising, Grünbaum observes, that they failed to worry about why there
is something rather than nothing. It was churchly philosophers like
Augustine and Aquinas who insinuated the idea into Western thought.
The doctrine of the world's ontological dependence on God—Grünbaum
calls it the Dependency Axiom—molded the intuitions of rationalists like
Descartes and Leibniz, predisposing them to believe that, were it not for
God's continuous activity of sustaining the world in existence, nothingness would prevail. Being without a cause was thus unthinkable to them.
Even today, when we ask why there is something rather than nothing at
all, we are, wittingly or not, heirs to a way of thinking that is a vestige of
early Judeo-Christianity.

So the Primordial Existential Question rests on the assumption of the
Spontaneity of Nothingness. The Spontaneity of Nothingness rests on the
Dependency Axiom. And the Dependency Axiom turns out to be a bit of
primitive and groundless theological bluster.

And that was only the beginning of Grünbaum's brief. He was not
content to observe that what he called the Primordial Existential Question rested on dubious premises. He wanted to show that these premises
were just plain *false*. There is no reason, in his view, to be astonished,
puzzled, awed, or mystified by the existence of the world. None of the
virtues claimed for Nothingness—its supposed simplicity, its natural-
ness, its lack of arbitrariness, and so on—made it the *de jure* favorite in
the reality sweepstakes: such was his conviction. In fact, if we look at the
matter empirically—the way modern, scientifically minded people ought

to—we'd find that the existence of a world is very much to be expected. As Grünbaum himself put it, "What could possibly be more commonplace empirically than that something or other does exist?"

Here was a man who thought *Why is there something rather than nothing?* was as much of a cheat as the question *When did you stop beating your wife?*

LATER THAT DAY, as I made my way across the bosky campus of the University of Pittsburgh toward my next rendezvous with Grünbaum, I was determined to champion the mystery of existence and the ontological claims of Nothingness. His office was atop the Cathedral of Learning—which, I was informed, was the tallest academic structure in the Western Hemisphere. It looked like an amputated and monstrously gigantified spire of a Gothic church. Entering the lobby, with its ribbed groin vaulting, I looked instinctively for a knave, an apse, an altar. But this was a secular cathedral, devoted not to the worship of some deity but to the pursuit of knowledge. All I saw instead was a bank of elevators. I took one of them to the twenty-fifth floor, where my mentor-turned-interlocutor was waiting for me.

After some small talk about psychoanalysis, I asked him whether he would be willing to concede that the concept of nothingness at least made sense. Isn't it possible that, instead of the world we see around us, there might have been nothing at all?

"That's something I've suffered through and worried about," he said, in his slow and deliberate diction. "People have made arguments against the coherence of the concept of nothingness, but many of those arguments seem fallacious to me. Take the claim that absolute nothingness is impossible because we can't picture it. Well, you can't picture hyper-dimensional physics either! But proving that the Null World is a genuine possibility is not my problem. It's the problem of Leibniz and Heidegger and Christian philosophers and all the boys who want to make hay out of the question *Why is there a world rather than nothing at all?* If nothingness is impossible, then, as the medievals used to say, *cadit quaestio*—'the question falls'—and I'll just go have a beer!"

But, I asked, isn't nothingness the simplest form reality could take? And wouldn't that make it the most expected way for reality to have turned out—unless, that is, there were some sort of cause or principle to fill the void with a world full of existing things?

"Oh, I'll grant that nothingness may be the simplest *conceptually*. But even if it is, why should this simplicity—this *presumed* simplicity—mandate the realization of the Null World in the absence of an overriding cause? What makes simplicity into an ontological imperative?"

It has become a "veritable mantra," Grünbaum complained, that the simplicity of nothingness makes it objectively more probable.

"Certain scientists and philosophers gawk at the world and say, 'We just know that simpler theories are more likely to be true.' But that's just their psychological baggage, their heuristic mode. It has nothing at all to do with the objective world. Look at chemistry. In ancient times, Thales held that all chemistry was based on a single element, water. When it comes to simplicity, Thales's theory wins hand over fist against Mendeleyev's nineteenth-century 'polychemistry,' which posits a whole periodic table of elements. But Mendeleyev's theory is the one that matches reality."

So I tried another tack. Simplicity apart, isn't nothingness the most *natural* form reality could have taken?

Grünbaum scowled slightly. "We know what's 'natural' only by looking at the empirical world," he said. "It's logically possible that a person might spontaneously metamorphose into an elephant, but we never observe such a thing. So we don't feel the slightest temptation to ask why this logical possibility is not realized. The collapse of a skyscraper, on the other hand, is something that *is* observed to happen from time to time. And when it does, we want an explanation, because it takes place against an empirical record of skyscraper collapse *not* occurring. Indeed, these nonoccurrences are so common that we are warranted in taking them to be 'natural.' When it comes to the universe, however, we've never observed its nonexistence, let alone found evidence that its nonexistence would be 'natural.' So why should we be tempted to ask for an explanation of why it exists?"

Here I thought I had him.

"But we *have* observed its nonexistence," I interjected. "The Big Bang theory tells us that the universe came into being only around 14 billion years ago. That's a drop in the bucket when you consider eternity. What was the universe doing in that infinite stretch of time before the Big Bang singularity, if not failing to exist? And wouldn't that make nonexistence its natural state?"

Grünbaum made short work of this objection.

"So what if the universe has a finite past?" he said. "Physics does not allow us to extrapolate back and say, 'Before this singularity there was nothingness.' That's an elementary mistake so many of my opponents make. They mentally picture themselves at the initial singularity as observers endowed with memory, and this gives them the irresistible feeling that there must have been earlier moments of time. But the lesson of the Big Bang model is that before the initial state there *was* no time."

Hmmm, I thought, Grünbaum seems to be a closet Leibnizian on the matter of time. In the late seventeenth century, Leibniz and Newton staked out competing positions on time's true nature. Newton took the "absolutist" position, holding that time transcended the physical world and all that went on within it. "Absolute, true and mathematical time, of itself and from its own nature, flows equably without relation to anything external," Newton declared. Leibniz took the opposite, "relationist" position. He argued, against Newton, that time was merely a relation among events. In a static world—a world without change, without "happenings"—time would simply not exist. Grünbaum, in contending that there was no time before the Big Bang, seemed to be echoing Leibniz. He was assuming that it would be meaningless to talk of time in a clockless and eventless state of Nothing.

But when I voiced this point, Grünbaum responded with a bit of jujitsu.

"No, Jim, I'm being philosophically elastic," he said. "I'm not necessarily siding with Leibniz. Maybe one can imagine time flowing in a Null World, as Newton did. But that's not how the Big Bang model works! The model itself says that the initial singularity marks a temporal boundary. If you take the model to be physically true, then that's where time begins."

So was he saying that the very idea of a world coming into existence out of nothingness was nonsensical?

"Yes, because it implies a process taking place *in time*. To ask how the universe came into existence in the first place presupposes that there were earlier moments of time when nothing at all existed. If the theory allowed us to talk about such earlier moments—time before the Big Bang—then we could ask what was going on then. But it doesn't. There is no 'before.' So there's no gap for God to sneak into. You might just as well say that the universe came out of nirvana!"

But it's not just religious believers who dwell on the gap between Nothingness and Being, I objected. Plenty of atheist philosophers are also on record professing astonishment that there should be a cosmos. I mentioned one in particular, J. J. C. "Jack" Smart—a tough-minded Australian philosopher of science and, like Grünbaum, an uncompromising materialist and atheist. Smart said that *Why does anything exist at all?* struck him as the "profoundest" of all questions.

"Well, I'll tell you something about Jack," Grünbaum replied. "He had a very religious upbringing. He may be an atheist now, but he once told me he'd be glad if someone could refute his arguments against religion, because he missed his old beliefs. People like him have a deep-seated tendency to be awed or amazed by the existence of the world. Like I say, they absorb it with their mother's milk."

I could not resist bringing up Ludwig Wittgenstein, who was also obsessed with the mystery of existence. Many philosophers deem Wittgenstein the greatest philosophical figure of the twentieth century. But Grünbaum, I quickly learned, was not among them.

"I'm sorry," he said, rolling his eyes, "but the paper where Wittgenstein talks about that is just *dreadful*. It's an unbelievably sick paper, semipsychotic. He gets to the end of his lecture and says he's in 'awe' of the question *Why is there something rather than nothing?* But he also claimed that the question had no sense! Then why is he still in awe of it if he'd debunked it? He needed to see a psychiatrist and not inflict his 'awe' on us."

I began to wonder whether Grünbaum might not be the most unflappable philosopher I had ever met. Clearly he did not suffer from any dread of Nothingness—what he derisively called the "ontopathological syndrome." Clearly he was unastonished by a world of Being. Did *anything* astonish the man? Was there *any* philosophical problem he found awesome and bewildering? What about, for example, the problem of how consciousness arises from brute matter?

"I'm amazed by the variety of consciousness and the kinds of things that the human mind can come up with," he said. "It's all very splendiferous! But I don't find the existence of consciousness puzzling."

I noted how different his attitude was from that of the philosopher Thomas Nagel, one of my intellectual heroes. In his book *The View from Nowhere*, Nagel pondered at length the mystery of how the mind's irreducibly subjective character could fit into the objective physical world.

"I've never read that book," Grünbaum said.

But it's such an important book! I stammered. The Oxford philosopher Derek Parfit declared Nagel's book the greatest philosophical work of the postwar era.

"Did he?" Grünbaum replied. "Well good for him! But as for me, why should *I* be puzzled that I'm put together the way I am? I know that many things have shaped my personal history. And there are many things about myself that I don't understand—why I have certain habits and tendencies, for example. But these are biological or bio-psychological questions. With enough evolutionary theory and genetics and what-have-you, they become potentially interesting. But I don't sit around wondering why I'm the way I am. I don't live in a limbo of dubiety."

If, as Aristotle remarked, philosophy begins with wonder, then it ends with Grünbaum.

Still, the scope of the man's knowledge was breathtaking. The nature of time, the ontological status of scientific laws, the extravagances of quantum cosmology: all yielded before his precise and rigorous understanding. And the sheer pleasure it all gave him ("I'm having a ball!") was contagious.

I asked him whether it was possible that an entity in our universe's distant future—an "omega point," as some thinkers have called it—might have reached back in time and retroactively caused the very Big Bang that brought the whole show into being.

"Ah," he said, "you're talking about retrocausation. Is such a thing possible?" He then launched into a learned disquisition on cause and effect whose virtuosity reminded me of a great diva delivering an opera aria. I listened with more awe than understanding as he wrapped it up: "Well, they got it wrong because they misextrapolated from second-order equations in Newtonian mechanics, where forces are causes of accelerations, to a third-order differential equation, Dirac's equation, in which forces are not causes of accelerations. So even though when you integrate over all future time you have force quantities in the integral—called 'pre-accelerations'—that doesn't mean that this instantiates retrocausation of acceleration by forces. Say, would you like a little gin? I think I've got some here."

As he reached into a lower desk drawer for the salutary bottle and a couple of glasses, I gratefully accepted the offer.

HAD GRÜNBAUM SHAKEN my conviction that the mystery I was pursuing was a genuine one?

Well, the Great Rejectionist had certainly changed my mind about one thing. Contrary to what I had assumed—along with just about every scientist and philosopher who has ever pondered the matter—the Big Bang does not, in itself, make the mystery of existence more acute. It does not mean that the cosmos somehow "leapt into being" out of a preexisting state of nothingness.

To see why, let's play the tape of the universe's history backward. With the expansion reversed, we see the contents of the universe coming together, growing more and more compressed. Ultimately, at the very beginning of cosmic history—which, for convenience, we'll label $t = 0$—everything is in a state of infinite compression, shrunk to a point: the "singularity." Now, Einstein's general theory of relativity tells us that shape of spacetime itself is determined by the way energy and matter are distributed. And when energy and matter are infinitely compressed, so too is spacetime. It simply disappears.

It is tempting to imagine the Big Bang to be like the beginning of a concert. You're seated for a while fiddling with your program, and then suddenly at $t = 0$ the music starts. But the analogy is mistaken. Unlike the beginning of a concert, the singularity at the beginning of the universe is not an event *in* time. Rather, it is a temporal boundary or edge. There are no moments of time "before" $t = 0$. So there was never a *time* when Nothingness prevailed. And there was no "coming into being"—at least not a temporal one. As Grünbaum is fond of saying, even though the universe is finite in age, it has *always* existed, if by "always" you mean at all instants of time.

If there was never a transition from Nothing to Something, there is no need to look for a cause, divine or otherwise, that brought the universe into existence. Nor, as Grünbaum observes, is there any need to worry about where all the matter and energy in the universe came from. There was no "sudden and fantastic" violation of the law of conservation of mass-energy at the Big Bang, as his theistically minded opponents have claimed. According to the Big Bang cosmology, the universe has always had the same mass-energy content, from $t = 0$ right up to the present.

Still, why should all of this matter and energy exist in the first place? Why do we find ourselves in a spacetime manifold with a certain geometrical shape and finite age? Why is this spacetime saturated with all kinds of physical fields and particles and forces? And why should those fields and particles and forces be governed by a particular set of laws, and a rather messy set at that? Wouldn't it be simpler if there were nothing at all?

Grünbaum had done his best to dispel the notion that there was anything metaphysically important about simplicity. He was willing to concede, for the sake of argument, that the Null World might well be the simplest form reality could take. Yet he could see no reason for this to stack the odds in favor of nothingness. "Why should we think that the *simple* is ontologically more likely to be *true*?" he kept asking rhetorically.

He had a point. And for some philosophers, this is where the argument grinds to a halt. Why should considerations of mere simplicity make us think that, barring some preternatural force or cause, there should be Nothing rather than Something? What's wrong, ontologically speaking, with complexity? Either you have a hunch that the sheer existence of the world needs an explanation, or you have a hunch that it doesn't. Grünbaum stood firmly in the latter camp, and no intuitions about the alleged simplicity of nothingness were going to move him.

But maybe he was undervaluing the power of simplicity. For scientists, after all, simplicity is nothing less than a guide to truth. As the physicist Richard Feynman put it, "The truth always turns out to be simpler than you thought." It is not that they want *reality* to be simple; rather, they want their *theories* of reality to be as simple as possible.

It's surprisingly tricky to say what makes one theory simpler than another. Still, there are some agreed-upon criteria. Simple theories posit few entities, and few *kinds* of entity; they obey the principle of Occam's razor: "do not multiply entities needlessly." Simple theories also have the minimum number of laws, and those laws take the simplest mathematical form. (Straight-line equations, for example, are deemed to be simpler than complicated curves.) Simple theories are also parsimonious when it comes to arbitrary features—unexplained numbers like Planck's constant and the speed of light.

Simple theories are obviously more convenient to use, more congenial to our intellects. They also appeal to our aesthetic sense. But why should they be more likely to be *true* than complex theories? This is a question that philosophers of science have never satisfactorily answered. "I suspect that it is not possible fully to justify the idea that simple theories are objectively more likely to be true than are complex ones," Jack Smart has observed. Nevertheless, when scientists have in hand two rival theories that are equally consistent with past evidence, it is the simpler of the two that they invariably prefer, since it is seen as more likely to be confirmed by future data. And the conviction that simpler theories are more probable than complicated ones is not confined to scientists. Suppose you have two equally well-confirmed theories, A and B. Theory A predicts that all life in the Southern Hemisphere will be wiped out tomorrow. Theory B predicts that all life in the Northern Hemisphere will be wiped out tomorrow. And suppose that Theory A is very complicated and Theory B is very simple. Then which of us northerners would not be trying to get on a plane for the Southern Hemisphere tonight?

If simple theories are indeed more likely to be true than complicated ones, that must be because the world as a whole has a deep-seated bias toward simplicity. Such a bias seems to have been successfully exploited by physicists in their search for the ultimate laws of nature. The "symmetries" that physicists look for in those laws are, as the Nobel laureate Steven Weinberg has pointed out, really principles of simplicity—principles that say, for example, the future should resemble the past in its most basic respects.

But simplicity is, for scientists, more than a guide to truth. It is also, as Weinberg has observed, "part of what we mean by an explanation." It is simplicity that distinguishes a "beautiful explanatory theory" in physics from a "mere list of data." Richard Dawkins made a similar point. Complicated realities, Dawkins submits, are more improbable than simple ones, and therefore stand in greater need of explanation. Take the existence of biological life. To posit a God as its cause is a nonstarter, Dawkins argues, since "any God capable of designing a universe, carefully and foresightfully tuned to lead to our evolution, must be a supremely complex and improbable entity who needs an even bigger explanation than the one he is supposed to provide." It is the simplicity of natural selection that makes it a satisfying explanation of life.

Now, the simplest theory of all is the one that says NOTHING EXISTS. This theory—the Theory of Nothingness—posits no laws and no entities; it has zero arbitrary features. If simplicity is indeed a mark of truth, then the Theory of Nothingness must enjoy the highest *a priori* probability. Absent any data about reality, the Null World is the one that should be expected to obtain. But it does not obtain! There is evidently a great abundance of Being. If we are scientifically minded, this should surprise us—shouldn't it?

Yet it did not surprise Grünbaum. So what, he said, if the Null World has the greatest *a priori* probability? "Probabilities just do not legislate ontologically," he kept insisting. Probability is not, in other words, a force driving the way reality should turn out, a force that had to be countered by *another* force, divine or otherwise, if there was to be Something rather than Nothing. That the universe seemed to confound the canons of science just didn't strike him as an intellectual problem.

Sometimes, of course, complicated theories do turn out to be true. As Grünbaum had pointed out, the modern theory of chemistry, which posits a whole periodic table full of elements, is a lot more complicated than the ancient chemical theory of Thales, based on water alone. But when scientists are faced with such complicated theories, they invariably look for simpler ones that underlie and explain them. A notable case is the contemporary quest for a unified theory of physics. Here the motive is to show that the four basic forces of physics—gravity, electromagnetism, the strong nuclear force, and the weak nuclear force—are all manifestations of a single underlying superforce. Such a unified theory—a "Theory of Everything," as it is sometimes called—would be superior to the partial theories it supersedes because of its relative simplicity. Instead of positing four forces, each governed by a distinct law, it would posit a single force-*cum*-law. In doing so, it would offer a more comprehensive explanation of nature than the current theoretical patchwork. Indeed, such a unified theory might turn out to be the closest we can come to giving a complete physical explanation of why the world is the way it is. But the final theory of physics would still leave a residue of mystery—why *this* force, why *this* law? It would not contain within itself an answer to the question of *why* it was the final theory. So it would not live up to the principle that every fact must have an explanation—the Principle of Sufficient Reason.

On the face of it, the only theory that does obey this principle is the Theory of Nothingness. That is why it's surprising that the Theory of Nothingness turns out to be false, that there is a world of Something. And any theory of this world of Something, however simple and ultimate, is doomed to fail the test of Sufficient Reason.

Or is it? Mightn't there be, after all, a theory of this world that leaves no unexplained explainers, one that reduces the residue of mystery all the way to zero? Finding such a theory would be tantamount to answering the question *Why is there something rather than nothing?* Adolf Grünbaum and his ilk might think this theory is not worth searching for—especially if the search takes a supernatural turn. But their arguments, while admittedly formidable, did not leave me convinced that the quest should be abandoned. There is nothing I dislike more than premature intellectual closure.

———

THAT NIGHT I got a personal glimpse into the Abyss of Nonbeing.

The plan for the evening seemed a good one. Adolf, accompanied by his wife Thelma, would pick me up at my hotel. Then we would set off for dinner at a restaurant called Le Mont, perched high above Pittsburgh on Mount Washington. The view was reputedly spectacular.

Adolf was driving a late-model Mercedes-Benz. His wife, a charming and somewhat abstracted woman of the same age, sat next to him. I sat, like their son, in the back seat.

It was when we got onto the freeway running along the Allegheny River that my pulse began to race. A diminutive man, shrunken by age, Adolf could barely see above the dashboard. It was like having, well, Mr. Magoo for a chauffeur. Oblivious to the heavy and fast-moving traffic around us, he maintained a constant monologue as he tried to work out the route. We were having one close call after another, but Adolf and his wife seemed blissfully unaware of the angry honks coming from the other cars. The longer we drove, the more Mount Washington seemed to recede from us. It was like a cruel real-life version of Zeno's paradox.

Eventually we somehow found ourselves on the *other* side of the mountain—where, perversely, the speed and volume of the traffic only increased. The angry honking around us continued, and the probability of escaping a serious collision seemed headed toward zero. Would I walk away from the smoking wreckage? Possibly: we were, after all, in a late-model Mercedes. But I couldn't help fearing that the precious flame of my consciousness was about to be extinguished eternally, that I was in danger of making the transition from Pittsburgh to Nothingness.

Finally Adolf responded to my frantic pleas to pull over with a breathtaking maneuver: he came to a dead stop in the middle lane. A passing state trooper took note of our predicament, and we were kindly set right and escorted to the mountaintop restaurant. On arriving, I found myself to be more than usually in need of a fortifying bumper of champagne.

"Go relax and enjoy yourself! Don't worry about why there's a world— it's an ill-conceived question!" Grünbaum exclaimed to me nonchalantly, with a trace of paternal affection, once the three of us had been seated at our table. The view was indeed stunning. All of Pittsburgh lay spread out below us. I could see where the Allegheny and the Monongahela came together to form the Ohio River. Bridges, festooned with twinkling lights, spanned the waters every which way.

The restaurant itself had a curiously 1950s feel to it, with older waiters in black tie, like extras in a Marx Brothers movie, and lots of crystal and brocade everywhere. Across the room, a local torch singer in sequins, accompanied by a pianist, belted out "At the Copa."

As I listened to my distinguished interlocutor above the music—"They need p and q, these boys, they need p and q!" he exclaimed, alluding to a pair of premises I had lost track of—a sort of metaphysical *tristesse* came over me. Earlier, on the road, I had had a near encounter with *le néant*. Now here I was in a provincial restaurant that, to a New Yorker like myself, seemed a vestige of a departed past, the snows of yesteryear. It was as if the Copa had never left Pittsburgh. In this eerily unreal setting, I could almost feel the Spontaneity of Nothingness. Okay, it was a mood, not a philosophical argument. But it filled me with the conviction that Grünbaum's ontological certitude—watertight, bulletproof, sunk-hinge, angle-iron, and steel-faced though it was—could not be the last word. The mystery of existence was still out there.

I was driven back to my hotel without incident. Slightly addled by the quantities of champagne and wine I had consumed, I lay down and drifted off to sleep without turning down the bedspread. The next thing I knew, the dawn light was filtering through the curtains and the phone was ringing. It was the Great Rejectionist.

"Did you sleep well?" he buoyantly asked.

5

FINITE OR INFINITE?

Compared to the eternal cosmos envisaged by the ancients, our own universe is something of a Johnny-come-lately. It seems to have been around a mere 14 billion years or so. And its future may well be bounded too. According to current cosmological scenarios, it is destined either to disappear abruptly in a Big Crunch some eons down the road, or to fade gradually into a dark and chill nothingness.

The temporal finitude of our universe—here today (but not yesterday), gone tomorrow—makes its existence seem all the more insecure and contingent. And mysterious. A world with solid ontological foundations, it seems, just wouldn't behave like this. It would exist eternally and imperishably. Such a world, unlike the finite Big Bang universe, would have an aura of self-sufficiency. It might even harbor the cause of its own being.

But what if our own world, contrary to current cosmological thinking, did turn out to be eternal? Would the mystery of its existence then become less acute? Or would the sense of mystery vanish entirely?

THE TEMPORAL NATURE of the world has long been a hotly contested issue in Western thought. Aristotle held the cosmos to be eternal, with no beginning or end in time. Islamic thinkers disagreed. The great philosopher and Sufi mystic al-Ghazālī, for instance, argued that the very idea of an infinite past was absurd. In the thirteenth century, the Catholic Church declared it to be an article of faith that the world had a beginning in time—although Saint Thomas Aquinas, showing some loyalty to the Aristotelian tradition, insisted that this could never be proved philosophically. Immanuel Kant argued that a beginning-less world led to paradox: how, he asked, could the present day ever have arrived if an infinite number of days had to pass first? Wittgenstein, too, felt there was something odd about the idea of an infinite past. Suppose, he said, you were to come across a man reciting to himself, "...9...5...1...4...1...3... finished!" Finished what? you ask him. "Oh," he says with relief, "I've been reciting all the digits of π backward from eternity, and I finally got to the end."

But is there anything truly paradoxical about an infinite past? Some thinkers object to the notion because it entails that an infinite series of tasks might have been completed before the present moment—which, they say, is impossible. But completing an infinite series of tasks is not impossible if you have an infinite amount of time in which to perform them all. In fact, it is mathematically possible to complete an infinite series of tasks in a *finite* amount of time, provided you perform them more and more quickly. Suppose you can accomplish the first task in an hour; then the second task takes you a half hour; the third takes you a quarter of an hour; the fourth takes you an eighth of an hour; and so on. At that rate, you will have finished the infinite series of tasks in a total of just two hours. In fact, every time you walk across a room you accomplish such a miracle—since, as the ancient philosopher Zeno of Elea observed, the distance you cover can be divided into an infinite number of tinier and tinier intervals.

So Kant and al-Ghazālī were wrong. There is nothing absurd about an infinite past. It is conceptually possible for there to have been an infinite succession of sunrises before the one this morning—provided there was an infinite span of time in which they could have occurred.

Scientific thinkers, by and large, have not shared such philosophical qualms about eternity. Neither Galileo nor Newton nor Einstein had any problem conceiving of a universe that was infinite in time. Indeed, Einstein added to his field equations a fudge factor—the infamous "cosmological constant"—to ensure that they would yield a universe that was static and eternal.

But astronomical observations soon revealed that, contrary to Einstein's intuition, the universe was *not* static. It was expanding, as if from an initial explosion. Even in the face of this evidence, some cosmologists still clung to the hope that the universe might be eternal. In the late 1940s, Thomas Gold, Hermann Bondi, and Fred Hoyle proposed a theoretical model called the "Steady-State Universe," which managed to be both expanding *and* eternal. (Gold and Bondi claimed that they came up with the idea after seeing the horror film *Dead of Night*, the dream-infused plot of which loops around on itself endlessly.) In their model, the empty space left behind by the ever-retreating galaxies is continuously filled by new particles of matter, which pop into existence spontaneously thanks to a "creation field." Thus despite the expansion, a constant density of matter is maintained. Even though it is constantly expanding, the Steady-State Universe always looks the same. It has no beginning and no end.

Another cosmological model of eternity is the "Oscillating Universe," which was first proposed in the 1920s by the Russian mathematician Alexander Friedmann. According to this model, our universe—the one that originated some 14 billion years ago with the Big Bang—emerged from the collapse of an earlier universe. And, like that earlier universe, ours too will eventually stop expanding and collapse back on itself. But when it does, the result will not be an all-annihilating Big Crunch. Instead, a new universe will rebound out of the fiery implosion, in what might be called the Big Bounce. And so on and so on, *ad infinitum*. In this model, time becomes an endless cycle of destruction and rebirth, rather like the dance of the god Shiva in Hindu cosmology.

Both the Steady-State Universe and the Oscillating Universe make the problem of cosmic origination go away. If the universe is infinitely old—if it has always been around, in other words—there is no "creation event" to be explained. Unfortunately for lovers of eternity, the Steady-State model is no longer taken seriously by cosmologists. It was done in by the detection, in 1965, of the background radiation left over from the Big Bang, which furnished decisive evidence that our universe had a fiery beginning after all.

The Oscillating model has fared better, but it is plagued by theoretical gaps. So far, no one has been able to explain exactly what sort of unknown repulsive force could overcome the attractive pull of gravity at the last moment of collapse and cause the universe to "bounce" rather than "crunch."

So at the moment, anyway, the odds seem to favor a finite past for our universe. But what if our universe is not all there is? What if it is a part of some greater ensemble?

One of the great lessons of the history of science is that reality always turns out to be more encompassing than anyone imagined. At the beginning of the twentieth century, our universe was thought to consist of just the Milky Way galaxy, sitting all by itself in an infinite space. Since then, we have learned that the Milky Way is merely one of a hundred billion or so similar galaxies. And that's just the *observable* universe. The current theory that best explains the Big Bang is called the "new inflationary cosmology." As it happens, this theory predicts that universe-engendering explosions like the Big Bang should be a fairly routine occurrence. (As one friend of mine observed, it would be very odd if the Big Bang came with a label that said, "THIS MECHANISM OPERATED ONLY ONCE.")

In the inflationary scenario, our universe—the one that suddenly popped into existence some 14 billion years ago—bubbled out of the spacetime of a preexisting universe. Instead of being all of physical reality, it's just an infinitesimal part of an ever-reproducing "multiverse." Although each of the bubble universes within this multiverse had a definite beginning in time, the entire self-replicating ensemble may be infinitely old. The eternity that seemed lost with the discovery of the Big Bang is thus regained.

With an eternal world—whether of the inflationary variety or some other—there is no inexplicable "creation moment." There is no role for a "first cause." There are no arbitrary "initial conditions." An eternal world thus seems to satisfy the Principle of Sufficient Reason. The way it is at any moment can be explained by the way it was the previous moment. Indeed, its *existence* at any moment can be explained by its existence the previous moment. Should that be enough to dispel any lingering sense of mystery?

Many have thought so—prominently among them, David Hume. In Hume's *Dialogues Concerning Natural Religion*, the character Cleanthes, who comes closest to being the author's mouthpiece, gives two arguments that an eternal world requires no explanation for its existence. "How," he asks, "can anything that exists from eternity have a cause, since that relation implies a priority in time and a beginning of existence?" It is assumed here that an explanation must invoke a cause, and that a cause must come before its effect. But nothing could precede a world with an infinite past, so such a world could have no prior cause and hence no possible explanation for its existence.

There are two problems with this first argument. To begin with, nothing in the concept of causation says that a cause must always precede in time its effect. Think of a locomotive pulling a caboose: the motion of the former causes the motion of the latter, yet the two are concurrent in time. Moreover, not all explanations must invoke causes. Think, for example, of the explanation for a rule in baseball or a move in chess.

Hume's second argument is a better one. Suppose (he has his spokesman Cleanthes say) we think of the history of the world as a series of events. If the world is eternal, this series is an infinite one, with no first or last member. Now, each event in the series can be causally explained by the event that precedes it. Since there is no event that lacks an explanation, everything seems to be explained. "Where then is the difficulty?" Cleanthes asks. He is unimpressed by the obvious rejoinder: that even if each event in the series is causally explained in terms of an earlier event, the series *as a whole* remains unexplained. For the series as a whole, he insists, is not something over and above the events of which it is composed. "I answer that the uniting of these parts into a whole, like the uniting of several distinct countries into one kingdom, or several distinct members into one body, is performed merely by an arbitrary act of the mind, and has no influence on the nature of things," Cleanthes says. Once all the parts are explained, he submits, it's unreasonable to demand a further explanation of the whole.

Seen in this light, an eternal world looks like the cause of itself, since everything within it is caused by something else within it. Hence it requires no external cause for its existence. It is *causa sui*—an attribute usually reserved for God.

But there's still something missing here. This infinite world is like a railroad train with an infinite number of carriages, each pulling the one behind it—and no locomotive. It can also be likened to a vertical chain with an infinite number of links. Each of these links holds up the link below it. But what holds up the chain as a whole?

Imagine yet another sort of series that has no beginning and no end, this one consisting of an infinite succession of copies of some book—say, the *Bhagavad Gita*. Suppose that each book in the series is faithfully copied by a scribe, letter for letter, from the preceding book in the series. Now, for each given copy of the *Bhagavad Gita*, the text is fully explained by the text of the preceding copy, from which it had been transcribed. But why should the whole series of books, extending back infinitely far in time, be copies of the *Bhagavad Gita*? Why not copies of some other book—*Don Quixote*, say, or *Paradise Lost*? Why, for that matter, should there be any books at all?

The preceding thought experiment, essentially due to Leibniz, is somewhat fanciful. But it can be sharpened up and made more scientific. Suppose you want to explain why the universe is the way it is at a given moment in its history. If the universe is eternal, you can always find earlier states in its history that are causally related to the state you're trying to explain. But knowledge of those earlier states is not enough. You must also know the *laws* that govern how one state of the universe evolves into another.

To be more precise, consider the total mass-energy of the universe as it is today. Call this mass-energy M. Why does M happen to have the value it does? To answer that question, you might point out that total mass-energy of the universe yesterday was also M. But that is not by itself an explanation of its value today. You also need to appeal to a law—in this case the law of mass-energy conservation. The total mass-energy of the universe today is M because (1) the total mass-energy of the universe yesterday was M and (2) mass-energy is neither created nor destroyed. Now your explanation is complete.

Or is it? It appears that there are *two* ways in which the universe could have been radically different. It could have had a different total mass-energy throughout its history—say, M' instead of M. And it could have had a different law governing that mass-energy: a law that, for example, might have allowed the mass-energy to cycle back and forth over time between a pair of values, M and M'. (To return to the *Bhagavad Gita* example for a moment, that would be as if the text kept getting translated back and forth from Sanskrit to English to Sanskrit to English and so on.) We still have no explanation of why there is *this* law and *this* precise value. Both appear to be contingent. Nor do we (yet) have an explanation of why there should be any mass-energy at all, let alone a law governing it. An eternal world can still be a mysterious world.

But we knew this intuitively already. Even if something is *causa sui*, its existence can still seem arbitrary. And an entity needn't be eternal to be self-caused. It could also trace out a circular path in time, looping around on itself so that it has no beginning and no end. Something of the sort can be found in the 1980 movie *Somewhere in Time*. The main character (played by Christopher Reeve) is given a gold watch by an old woman. He then travels back in time and gives the watch to the same woman when she was in her youth—the very watch that she will, some decades later, give to him. How did this watch come into being? In its entire existence, which spans only a few decades, it never sees the inside of a watch factory. It exists even though it has no creator. It seems to be *causa sui*. (Some physicists call an entity with such a circular history a *jinn*, since, like Aladdin's genie, it seems to be self-conjuring.) The existence of this gold watch is as inexplicable as the existence of the poem "Kubla Khan" would be if I had gone back in time to the autumn of 1797 and dictated it to a grateful Coleridge, who then published it so that two centuries later I could learn it by heart.

Could anything be more of an affront to the Principle of Sufficient Reason than a self-composing poem or a self-conjuring watch? Could anything be less self-explanatory than an Oscillating Universe, eternally bellowing in and out like some cosmic accordion, or an Inflationary Multiverse, endlessly frothing away like a just-uncorked bottle of Veuve Clicquot? Why such an absurdly busy cosmos? Why *any* cosmos at all, whether finite or infinite?

Why not nothing?

Night Thoughts at the Café de Flore

"*Et pour vous, monsieur? Du café? Une infusion?*"

The waiter posed the question in a tone of weary impatience. It was, after all, nearly closing time at the Café de Flore, on a late-winter night in Paris. The evening had been a hefty one, and I felt I needed something more fortifying than the options proposed. My companion, an aging but handsome voluptuary named Jimmy Douglas, suggested, as an alternative, a strongly alcoholic herbal concoction I had never heard of. It would, he insisted, buck up my liver.

It certainly seemed to have worked for him. Despite a life of riotous excess and free indulgence of his voracious and irregular appetites, Jimmy had remained preternaturally youthful. Friends called him Dorian Gray. (It perhaps helped that, as an heir to the Quaker Oats fortune, he did not have to toil for a living.) In the 1950s, he was the paramour of Barbara "poor little rich girl" Hutton, taking up with her after her fifty-three-day marriage to the international playboy/diplomat/polo-star Porfirio Rubirosa (a tough act to follow). In the 1960s, Jimmy threw a joint party for the Beatles and the Rolling Stones in his grand apartment in the Faubourg Saint-Germain, which adjoined that of a former French prime minister. Now, decades later, he was regaling me with stories of Baron Gottfried von Cramm and Nancy Mitford and the Aga Khan, and urging me to decamp from New York to Paris, where, he claimed, the nightclubs were better and the bacterial flora kept one eternally young.

Sipping the bracingly pungent herbal stuff the waiter had brought me, I looked around the Flore. At that hour, the café was hardly the "fullness of being" described by Sartre. At a table in the back I spotted Karl Lagerfeld, with his characteristic ponytail, dark glasses, and high white collar, in hushed conversation with one of his muses, who was wearing what looked like black lipstick. Other than that, the place was pretty much empty: *le Néant*.

But then there was a noisy burst of activity. A woman of a certain age, evidently an old friend of Jimmy's, breezed through the front door, accompanied by a pair of what appeared to be Cuban gigolos dressed in shell suits. Giggling and grinding their teeth, this trio sat down with us and began to jabber away. The woman's face was a sallow mask of leathery jollity, and she talked in a low croak that put me in mind of Jeanne Moreau. I listened with a kind of ironic inattention, but my spirits began to flag.

It seemed a good time to leave.

The late-night air was chill and damp. As I started to walk back to my hotel, I glanced across the deserted square at the Église de Saint-Germain-des-Prés, built a thousand years ago. There, in one of the side chapels, reposed the body of Descartes. (Well, most of it, anyway—the whereabouts of his skull and right forefinger is a mystery.)

I wondered if Sartre, scribbling inside the Café de Flore, used to feel the Cartesian presence from across the square. And Descartes wasn't the only philosophical specter lurking about. Directly across the Boulevard Saint-Germain from the café is the rue Gozlin, which runs for a single block. It is the last vestige of the rue Sainte-Marguérite, a medieval street that was absorbed into the boulevard during Baron Haussmann's modernization of Paris in the mid-nineteenth century. There, some centuries ago, stood the Hôtel des Romains, where Leibniz lived for two of the very happy four years of his life he spent in Paris.

What was Leibniz doing in Paris? As usual with him, intrigue lay behind his visit. He had come to the French capital in 1672 on a secret diplomatic mission to persuade Louis XIV to invade infidel Egypt rather than Christian Germany. The mission was not a success. "As to the project of Holy War," the Sun King was said to have politely responded to Leibniz, "you know that since the days of Louis the Pious such expeditions have gone out of fashion." (In the event, France invaded Holland.)

But Leibniz's time in Paris was hardly wasted. It was while staying in the Hôtel des Romains, in his thirtieth year—something of an *annus mirabilis* for him—that he invented the calculus (including the *dx* notation and the elongated "S" symbol for the integral that are in universal use today). And it was at that hotel, in his room overlooking the present site of the Café de Flore, that Leibniz began to lay the foundations for his later metaphysical philosophy, which would culminate with the posing of the deepest of all questions: *Why is there something rather than nothing?*

Both Leibniz and Descartes, in their rationalist way, confronted the mystery of existence. Both decided that the one sure ontological foundation for a contingent world like ours was an entity that carried within itself the logical guarantee of its own existence. Such an entity, they held, could only be God.

Like his philosophical forebears, Sartre too was a rationalist. Unlike them, he thought that the very idea of God was shot through with contradictions. Either a being has consciousness or it does not. If it does, it is *pour soi* ("for itself"), an activity rather than a thing, a "wind blowing from nowhere toward the world." If it does not have consciousness, it is *en soi* ("in itself"), an object fixed and complete. God, were such a being to exist, would have to be both *pour soi* and *en soi*: both conscious and complete in himself. And that, submitted Sartre, is an impossibility. Still, this God-like combination of fluidity and fixity is one we humans cannot help aspiring to. Our desire to be radically free and yet absolutely secure in our identities is, for Sartre, nothing less than a desire to be God. It is *mauvaise foi* ("bad faith"), a sort of original sin. It was what, according to Sartre, my waiter at the Café de Flore was displaying. "His movement is quick and forward, a little too precise, a little too rapid. . . . He bends forward a little too eagerly; his voice, his eyes express an interest a little too solicitous for the order of the customer. . . . He is playing, he is amusing himself. But what is he playing? We need not watch long before we can explain it: he is playing at *being* a waiter in a café." But a consciousness can never have an essence, like waiterhood or divinity. Thus God is a conceptual absurdity. And man is "a useless passion."

Such Sartrean reflections engrossed me as I set out on my nocturnal walk home—past the elegantly illumined Théâtre de l'Odéon, around the edge of the Jardin du Luxembourg, and then toward my hotel in Montparnasse— which was not far, as it happened, from the cemetery where Sartre and Simone de Beauvoir are buried (Susan Sontag too). The quiet that comes over Paris in the small hours—on some streets you can even hear the echo of your footfall, which is unthinkable in New York—made my thoughts seem clear and compelling and true.

The next morning, however, a metaphysical fog had descended on me again. I wondered whether there wasn't something unwholesome about the Café de Flore. Sartre's paradoxes seemed too easy to me, his ontological despair slightly off key. After all, Leibniz and Descartes were far greater philosophers than he ever was. And both of them were convinced that the world of contingent being—the one Sartre found so gooey and absurd, so permeated with nothingness—must rest on a secure and necessary onto- logical foundation.

There must be serious thinkers who still believed this. But I wasn't going to find them easily on the Left Bank, not, at least, in this century. Better to look for enlightenment in a more cloistral, medieval setting. So, after grabbing a *tartine et café crème* at the bar of Le Select, I hauled my bags onto the Métro and headed to the Gare du Nord, there to catch the Eurostar train to London. Arriving at Waterloo Station a few hours later, I caught the tube to Paddington, where I hopped on a local train to Oxford, debouching from the station into that city of dreaming spires well in advance of cocktail hour.

"I HAVE BEEN here before," I thought to myself (rather derivatively) as I made my way down Oxford's High Street. And I had—for the wedding of a friend just a few months earlier. Now it was midwinter, Hilary Term, and the clear light of the late afternoon leant an apricot glow to the Cotswold sandstone of Oxford's colleges. Bells rang out over the gables, cupolas, and finials. Students hurried to and fro through the Gothic labyrinth of passageways, cloisters, alleys, and quadrangles. All around me I felt the soft breath of a thousand years of learning.

So much for the bogus poetry. Where was the next clue to the mystery of the world's existence?

I had a pretty good idea. Years ago, in a stack of galleys I had been sent to review, one slim volume stood out. Its title, *Is There a God?*, was not in itself remarkable. Books with titles like that are a dime a dozen. What struck me were the credentials of the author, whose name was Richard Swinburne. He was a philosopher of religion, a practitioner of what is called "natural theology." But he was also a philosopher of science, the author of rigorous treatises on space, time, and causality. And he was clearly a thinker alert to the mystery of existence. "It is extraordinary that there should exist anything at all," I read on the back cover of the volume. "Surely the most natural state of affairs is simply nothing: no universe, no God, nothing. But there is something. And so many things. Maybe chance could have thrown up the odd electron. But *so* many particles!" What could account for the existence of such a rich and plenitudinous universe? And what could account for its many surprising features—notably its spatial and temporal order, its fine-tuned fostering of life and consciousness, its suitability as a theater for human action? "There is a complexity, particularity, and finitude about the universe that cries out for explanation," he wrote.

The simplest hypothesis that explains the existence of such a world is the hypothesis that God is behind it—that was Swinburne's conclusion. Admittedly, it was not a very original one. What was original was Swinburne's methodology. He did not pretend to *prove* God's existence by means of an abstract logical deduction, in the manner of Anselm, Aquinas, or Descartes. Instead, he used modern scientific reasoning. He endeavored to show that the God hypothesis was at least probable, more probable than its negation, and hence that belief in God was rational. "The very same criteria which scientists use to reach their own theories lead us to move beyond those theories to a creator God who sustains everything in existence," Swinburne wrote. Each step in his case was painstakingly justified by appeal to the canons of inductive logic. He was especially expert in the use of "Bayes's theorem," a mathematical formula that describes how new evidence raises or lowers the probability of a hypothesis. Using Bayesian confirmation theory, Swinburne sought to show that on the total evidence—which includes not just the existence of the universe, but also its lawfulness, the patterns of its history, and even the presence of evil within it—it was more likely than not that there is a God. Intellectually, this struck me as a bravura performance. Yet I knew that it did not strike everyone that way. Swinburne's fellow philosopher of science, Adolf Grünbaum, had been withering in his scorn for Swinburne's pro-theist case, calling it "a very poor job." Swinburne's reasoning in behalf of theism was "unsound" and "defective," Grünbaum had told me, full of "red herrings" and "strawmen." Over the years, Swinburne and Grünbaum have tussled repeatedly in forums like the *British Journal for the Philosophy of Science*. When I went back and read their exchanges, it was like witnessing a fiendishly intricate metaphysical Ping-Pong match. "Why, oh why," Grünbaum irritably asked at one point, "does Swinburne reason with Leibniz that even the bare of existence of the universe imperatively calls for a 'cause acting from outside'?"

Richard Dawkins was also skeptical, to say the least. In *The God Delusion*, Dawkins mocked Swinburne's claim that the God hypothesis possesses the scientific virtue of simplicity, calling his reasoning "a breathtaking piece of intellectual chutzpah." How, Dawkins asked, could a being who created and sustained a complex universe like our own, a being supposedly capable of monitoring the thoughts of all his creatures and answering their prayers ("Such bandwidth!"), be *simple*? As for Swinburne's argument that the existence of an omnipotent and infinitely loving God could be squared with a world containing evil and suffering, Dawkins deemed it "beyond satire." He recalled a televised discussion in which Swinburne (in Dawkins's words) "attempted to justify the Holocaust on the grounds that it gave the Jews a wonderful opportunity to be courageous and noble"—at which point a fellow panelist, the Cambridge chemist and arch anti-theist Peter Atkins, growled at Swinburne, "May you rot in hell."

A man capable of producing such bold reasoning about the cosmos, and of evoking such acidulated reactions among his foes, was clearly a man worth talking to. Swinburne had recently retired from Oxford, where he had been the Nolloth Professor of the Philosophy of the Christian Religion and a fellow of Oriel College. When I managed to get in touch with him, he was the soul of kindness, inviting me to come over to his residence in North Oxford for tea and a chat.

So the next afternoon I left my hotel on the High Street, made my way down Queens Lane, passed under the Bridge of Sighs and by the Bodleian Library and Ashmolean Museum, and emerged finally onto the broad Woodstock Road, which I followed for a mile or two to North Oxford. I chanced to notice an Eastern Orthodox Church as I turned off the main road to seek the address Swinburne had given me, which proved to be that of a 1950s modernist apartment block bordered by a row of handsome Edwardian brick houses. The still winter air of the neighborhood was improbably rich with birdsong. It seemed a good portent.

6

THE INDUCTIVE THEIST OF NORTH OXFORD

"You've come a long way," said Richard Swinburne as he welcomed me at his doorstep. Yes, I thought to myself, I have—all the way from the Café de Flore of post-Sartrean Paris to a philosopher-monk's cell in medieval Oxford.

Swinburne, born in 1934, was lithe and youthful-looking for a man in his mid-seventies. He had pleasant, rather clerical features, and a serene manner. His forehead was high and narrow, topped by a full head of gray hair. He spoke in a quiet voice, with a slightly nasal timbre, precise vowels, and an infinity of subtle modulations. He was wearing a nicely tailored dark suit and a sweater, which was tucked into his pants.

Swinburne, I discovered, lived alone in his cozily austere duplex apartment. We walked up a narrow flight of stairs to his study, where a crucifix hung on the wall. He absented himself for a moment and returned with a pot of tea and a plate of sugar biscuits.

I mentioned the interesting day I had spent with his great cosmological adversary, Adolf Grünbaum, and how dismissive Grünbaum had been of Swinburne's beliefs—in particular, his conviction that the sheer existence of the world cried out for some kind of explanation.

"Grünbaum misunderstands me," he responded mildly, in the manner of a curate discussing a difficult rector. "He represents me as saying that reality ought to be geared to throw up Nothing, and that it is unusual and surprising that it has thrown up Something. But that's not my position. My position is based on an epistemological principle: that the simplest explanation is most likely to be *true*."

And why, I asked, is simplicity such an epistemic virtue?

"There are innumerable examples to illustrate this," he said, "and not just from science. A crime has been committed. A bank has been robbed. There are three clues. A chap called Jones was reported to be near the scene of the crime at the time of the robbery. Jones's fingerprints were found on the safe. Money from a bank robbery was found in Jones's garret. Plausible explanation: Jones did the crime. Why do we think that? Well, if the hypothesis that Jones did the crime was true, you would probably find such clues; and if it wasn't, you probably wouldn't. But there are an infinite number of other hypotheses that meet this dual condition—for example, the hypothesis that somebody dressed up like Jones as a joke and happened to walk near the bank; and another person, not in collusion with the first, had a grudge against Jones and put Jones's fingerprints on the safe; and a third person, having no connection to the previous two, put the proceeds from a quite different robbery in Jones's garret. That hypothesis also meets the dual condition for being true. But we wouldn't think much of any lawyer who put it forward. Why? Because the first hypothesis is *simpler*. Science always reaches for the simplest hypothesis. If it didn't, one could never move beyond the data. To abandon the principle of simplicity would be to abandon all reasoning about the external world."

He looked at me gravely for a moment and then said, "Would you like some more tea?"

I nodded. He refilled my cup.

"Descriptions of reality can be arranged in order of their simplicity," Swinburne continued. "On *a priori* grounds, a simple universe is more likely than a complicated one. And the simplest universe of all is the one that contains *nothing*—no objects, no properties, no relations. So, prior to the evidence, that is the hypothesis with the greatest probability: the hypothesis that says there is Nothing rather than Something."

But simplicity, I said, did not force this hypothesis to be true. I refuted it by holding up a sugar biscuit.

"Right," said Swinburne, "so the question is, what is the simplest universe that contains the sugar biscuit and the teapot and us and everything else we observe? And my claim is that the simplest hypothesis explaining it all is the one that posits God."

The notion that there's anything simple about the God hypothesis is one that drives a lot of atheist thinkers—Richard Dawkins, for example—up a wall. So I had to challenge Swinburne on that. First, though, a slightly less fraught subject: did it matter to his case for God whether the universe had a finite or an infinite past?

"I know that a lot of thinkers look at the Big Bang through metaphysical spectacles," he said. "But I don't think the issue of a cosmic beginning is deeply relevant. Nor did Aquinas. Aquinas thought that, as far as philosophy was concerned, the universe might well have been infinitely old. It was a matter of Christian revelation that it came into existence at a particular moment in time. That's one way of reading Genesis. But suppose the universe has been going on forever, and that it's always been governed by the same laws. It remains true that there *is* a universe, and there might *not* have been. Whether the laws that govern its evolution have been in operation for a finite or an infinite time, they're still the same datum. And, for those laws to give rise to humans, they have to be of a very special sort. You might think that, given an infinite amount of time, matter will rearrange itself sufficiently to produce conscious beings. But that's not so! Think of the balls careening around on a billiard table. Even in an infinite time, they will not assume all possible configurations. A cosmos must meet some very precise conditions in order for humans to appear."

But what if our world is just one among a vast multitude of universes, each with different laws? Wouldn't some of them be bound to produce beings like us?

"Yes, I know that the multiple-universe idea has captured a lot of headlines," he said. "But that's not relevant to my case either. Suppose each universe throws off daughter universes that differ from the mother universe in various ways. How can we know such daughter universes exist? Only by studying our own universe and extrapolating backwards and finding that, at some point, another universe must have split off from it. Our sole source of knowledge about *other* universes is a detailed study of *this* universe and its laws. How then can we suppose that those other universes are governed by totally *different* laws?"

Perhaps, I said, the laws governing the other universes were the same, but the "constants" that occurred in those laws—the list of twenty or so numbers that determine the relative strength of the physical forces, the relative masses of elementary particles, and so forth—differed from one universe to the next. If our universe is but one among a vast ensemble of universes in which such constants varied at random, then isn't it to be expected that some of these universes should have the right mix of constants for life to occur? And, as humans, wouldn't we be bound to observe ourselves living in one of the universes whose features happened to be congenial to our existence? Doesn't this "anthropic principle" make the apparent fine-tuning of our universe wholly unremarkable? And, in that case, wouldn't the God hypothesis be unnecessary as an explanation of why we are here?

"Right," he said, with a faintly audible chuckle, as though he had heard this very point innumerable times before. "But then we would need to find a *law* of how these constants varied from universe to universe. If the simplest theory is one where the constants of nature undergo some change when a mother universe gives birth to a daughter universe, then that raises the question of why the multiverse is like that, as opposed to all the infinite *other* ways a multiverse might be. Those other multiverses would *not* give rise to universes with life. In any case, to posit a trillion trillion other universes to explain the life-fostering features of our universe seems slightly mad when the much simpler hypothesis of God is available."

But is the God hypothesis really all that simple? There is, I was willing to concede, a sense in which God might be the simplest thing imaginable. The God of the theologians is defined as the entity—or "substance," to use the technical term—that possesses every positive attribute to an *infinite* degree. He is infinitely powerful, infinitely knowledgeable, infinitely good, infinitely free, eternally existing, and so forth. Setting all the parameters equal to infinity makes a thing easy to define. In the case of a finite being, by contrast, you have to say it's of such-and-such size and has such-and-such degree of power, that it knows this much and no more, that it began to exist at such-and-such time in the past, and so on. In other words, there is a long and messy set of finite numbers to specify.

Now, in science, infinity is a very nice number, along with its opposite, zero. Neither infinity nor zero needs an explanation. Finite numbers do need explanations, however. If the number 2.7 occurs in your equation, someone will always ask, "Why 2.7? Why not 2.8?" The simplicity of zero and infinity precludes such awkward inquiries. The same logic might be said to apply to God. If the cosmic creator could make a universe only of such-and-such mass, but no heavier, then the question would arise of why there was such a constraint on its power. With an infinite God, there are no such limits to be explained.

So the God hypothesis does possess a certain sort of simplicity. But Swinburne's God is not mere infinite substance. He also intervenes in human history. He answers prayers, reveals truths, causes miracles to occur. He even incarnated himself in human form. This is a God that acts with complex purposes. Doesn't the ability to act according to complex purposes imply a corresponding complexity within the agent? Swinburne himself, I had noticed, seemed to assume as much in some of his writings. For instance, in a 1989 essay he observed that we humans could have complex beliefs and purposes only because we had complex brains. Wouldn't God, in order to accomplish what he does, have to be internally complex on a much vaster scale—infinitely complex, in fact?

Swinburne knit his tall brow a bit when I asked the question. But in an instant, it was unknit again.

"Humans need bodies if they are to interact with the world and benefit one another," he said. "And that necessitates having a complicated brain. But God doesn't need a body or a brain. He acts on the world directly."

But, I objected, if God created the world for a purpose, if he has complicated designs for his creatures, then his mind must contain complicated thoughts. So the divine "brain," even if it is wholly immaterial, must still be a complex medium of representation, mustn't it?

"It isn't logically necessary to have a brain of any kind to have beliefs and purposes," Swinburne replied. "God can see all of creation without a brain."

Wouldn't the ability to see all of creation, brainlessly or not, imply something other than simplicity? If God possessed within himself all knowledge of the world, wouldn't his internal complexity have to be at least equal to that of the world?

"Hmmmm," Swinburne said, stroking his chin. "I see what you're getting at. But look, there are all sorts of things I can do—tie my shoelaces, for example—without thinking about how I do them."

Yes, I said, but you can tie your shoelaces only because you have complicated neural circuits in your brain.

"That is, of course, true. But it's one truth that I can tie my shoelaces without thinking. It's another truth that there are certain things going on in my brain. These are two truths about the world, and they are not necessarily connected with one another."

I wanted to protest this weird mind-body parallelism he seemed to be buying into, this idea that mental processes and brain processes somehow stream along independently of each other. But I was afraid I was beginning to bore him.

"Let me put the point slightly differently," Swinburne said, "by way of an analogy. Someone like Dawkins might claim that science never posits the kind of 'omni' properties—omni-knowledge, omni-power—that we ascribe to God. But let's look at Newton's theory of gravitation. This theory postulates that every particle in the universe has one power and one liability. The power is to exert gravitational force, and the liability is to be subject to it. And the power is an *infinite* one: each particle influences every other particle in the universe, no matter how far away. So serious physicists have attributed an infinite power to very tiny particles. It's considered quite proper in science to attribute omni-properties to very simple kinds of objects."

We had apparently reached an impasse on the issue of simplicity. So I tried to find another weak point in Swinburne's case.

"It seems to me that your God is closer to an abstract ontological principle than to the heavenly father-figure that religious believers pray to," I said. "There may be, as you say, a supremely simple entity that explains the existence and nature of the universe. And it may even have some personal characteristics. But to equate this entity to the one that is worshipped in churches seems a bit far-fetched. It's easy to see how today's religions grew out of animistic cults and then got more sophisticated, as magical notions of the world gave way to scientific understanding. But those primitive cults weren't hooking on to anything transcendental."

"I think that's wrong," Swinburne said, quite abruptly and with some severity. "I think it's *always* been a matter of the transcendental. The God written about in the New Testament and some of the Old Testament is an omnipotent, omniscient, and all-good creator. And going right back to Jeremiah, you have the idea that the visible world holds evidence of the transcendental. Jeremiah talked of the 'covenant of night and day' that God has made. What this means is that the regular alternation of night and day shows the *reliability* of the creator. And this is, in essence, what philosophers call the argument from design—one of the central arguments for the existence of God. The early Christian, Jewish, and Islamic traditions all have this kind of transcendental thinking in their background. They just don't talk about it a lot, because the issue back then was not whether there is a God, but what he was like and what he had done."

Why, though, should someone who did not grow up in one of these traditions believe in such a God, one who cares about our actions and fates? Why not the abstract and aloof God of the eighteenth-century deists, or the impersonal God of Spinoza?

"Well," Swinburne said, "those conceptions fail to take seriously the infinite goodness of the creator. Now, what would a good God do? It's unlikely he'd create a universe and then not take an interest in it. Parents who leave their children to fend for themselves aren't very good parents. You'd expect God to keep a connection with his creation, and if things go wrong, to help people to straighten them out. He will want to interact with his creation, but not be too obvious about it. Like a good parent, he'll be torn between interfering too much and interfering too little. He'll want people to work out their own destiny, to work out what is right and wrong and so on, without his intervening all the time. So he'll keep his distance. But on the other hand, when there has been a lot of sin around, he will want to help people deal with it, especially those who *want* his help. He'll hear their prayers and sometimes he'll answer them."

I mentioned the argument of some philosophers that the universe was brought into existence not by a personal God, but by an abstract principle of goodness. That, after all, was what Plato believed.

"Philosophically, the idea of a Platonic principle of goodness is highly suspect," he said. "But I have a particularly Christian problem with it. Such an abstract principle can't deal with the problem of evil. There is, as we know, evil and suffering in the world. I have a theodicy—a view of why God should allow evil to happen. I think he allows it to happen because it's logically necessary if certain goods are to be possible, the goods arising from our possession of free will. God is omnipotent. He can do anything that is logically possible to do. And it isn't logically possible for him to give us free will and yet to ensure that we always use it in the right way."

Swinburne paused to take a sip of tea. When he began talking again, his tone had grown almost homiletic. "Now, a good parent allows his children to suffer, sometimes for their own good, and sometimes for the good of other children. A parent who does that, I think, has an obligation to *share* the child's suffering. Here's an example, perhaps a superficial one. Suppose my child needs a special medicine that's in short supply. I happen to have plenty of that medicine for my child. But suppose my neighbor's child suffers from the same disease and also needs that medicine. If I share my supply with my neighbor, my own child will have just enough of the medicine to survive. It's generally believed that it would be okay to make my child suffer so that the other child would survive too. But if I do this, I think I have an obligation to share my child's suffering. And God has the same sort of obligation. If he makes us suffer for a good cause, there comes a point where he has an obligation to suffer *with* us. And an abstract principle of goodness can't do that."

Despite the gravity of his point, I detected a quaver of mirth in Swinburne's voice, as if he was pleased by this intellectual twist.

"There's also the Christian doctrine of the atonement," he went on. "If my children do bad things to one another, they're wronging me too, because I've lavished a lot of care in trying to prevent this from happening. So, in wronging one another, we wrong God too. What's God going to do about that? Well, what do *we* do when we've wronged somebody? We make atonement. And there are four elements of atonement: repentance, apology, reparations, and penance. Humans have wronged God mainly by living the wrong sort of life. So how are we going to make it up? Well, we don't have much time—or inclination—to lead perfect lives, so we can't really make adequate reparations. On the other hand, making reparations is something that somebody else can help you with if you're not in the position to do it. In the Christian account, Jesus lived the perfect life, the one that we should have lived. And even though we have lived bad lives, we can offer Jesus's life in reparation for our own failings. In doing so, we show God that we take those failings seriously, so he will forgive us. That's the Christian doctrine of the atonement—part Aquinas, part Anselm. It follows from the nature of goodness itself that God will get involved in his creation. That's a sort of bridge between philosophy and Christianity."

There was something numinous in his logic. The question *Why is there something rather than nothing?* had led this philosopher not just to God, but all the way to the historical person of Jesus Christ.

I became aware again of the crucifix hanging on the wall just behind him. Was Swinburne a Roman Catholic? Or was he a member of the Church of England?

"Neither," he said. "I'm Eastern Orthodox."

"Oh," I blurted out, finding myself at a loss for anything to say.

But Swinburne turned out not to be orthodox in every sense. When I resumed the conversation, I raised the generally accepted theological axiom that God stands outside of time, apprehending the entire history of the cosmos at a glance from the unchanging perspective of eternity. Scholastic thinkers like Aquinas held that such timelessness was one of God's perfections.

"I don't endorse that view," he said, "and I don't think the Biblical writers did either. They thought of God as being within time, and I do too. The idea that there's a *before* and an *after* for God, that there's a sense to saying, 'He did this first and then that,' is coming back into fashion."

Why, I wondered aloud, did philosophers of religion so often fail to agree on such fundamental matters? And why was there such a vast metaphysical gulf between Swinburne, who thought the God hypothesis furnished a scientifically viable explanation for the existence of the world, and philosophers like Grünbaum, for whom the very idea was absurd?

"That in itself is an interesting question," Swinburne said. "And it's not confined to the philosophy of religion. You find such radical disagreement in every branch of philosophy you can name. And it can have practical consequences. People change their views about the morality of war, of capital punishment, a whole range of moral issues, based on philosophical arguments. But philosophy is a *terribly difficult* subject, and sorting out the hardest questions in the finite time of a human life is asking a lot. And we're not only finite, we're imperfectly rational. Our prejudices creep into our philosophical thinking, especially when it touches on our lives. They cause us to look at certain arguments more carefully, more sensitively, and perhaps to overlook others. Many philosophers were brought up in strictly religious households. As adolescents, they found their religion in conflict with things that were obviously true, and they rebelled against it. Then later, when someone shows them a more appealing sort of religion, they're not going to grasp it."

For Swinburne, God was not only a supernatural being to be worshipped and obeyed, but also the terminus of an explanatory chain. One could go no further than God in the quest to resolve the mystery of existence. Swinburne was not a believer in the Principle of Sufficient Reason. He did not think there was an explanation for *everything*. The metaphysical task, as he saw it, was to find the right stopping point in explaining the world, the one that would minimize the part of reality that was left unexplained. And that stopping point should be the simplest hypothesis that can encompass all the evidence before us.

Still, I could not resist raising the question of why God himself exists. Swinburne had conceded that the "most natural" state of affairs was absolute nothingness: no universe, and no God either. He also thought that a reality consisting of a universe and no God—the kind of reality that atheists believe in—was at least conceivable. Here Swinburne was at odds with many of his theological allies. From Anselm to Descartes to Leibniz on down to present-day philosophical theists (like Alvin Plantinga, of Notre Dame), they have viewed God's existence as a matter of necessity. Unlike our contingent universe, they held, God could not fail to exist; he contains within himself his own sufficient reason. Indeed, they insisted, his existence could be proved as a matter of logic. Swinburne dissented on this point. Where other philosophical theists talked of *necessity*, he talked of *simplicity*; and simplicity, as he saw it, made a hypothesis only *probable*, not undeniably certain. One could gainsay God's existence, he held, without being convicted of illogic.

But would Swinburne go so far as to say that God's existence was a "brute fact"?

"Yes, I would," he replied. "I would say that. It's not merely that there is no explanation for God's existence. There *couldn't* be an explanation. One of God's properties is omnipotence. If anything happens to him, it's because he allows it to happen. Therefore, if something else brought about God, it could only be because God *allowed* it to bring about God."

Here was a line of reasoning I certainly hadn't heard before. "So you're not personally puzzled," I said, "about why God exists—or, I don't know, maybe you *are* puzzled."

Swinburne chuckled—out loud, for once—and said, "I don't think that *anyone* thought that God was a logically necessary being, not at least until Anselm came along with his ontological proof. And that's halfway through the two millennia of Christianity. Anselm's ontological argument was a bad, unnecessary turn for theology. Even Aquinas didn't really believe in it. So I'm not alone in thinking that God does not exist as a matter of pure logic. But I do think that God is a necessary being in the sense that he does not *depend* for his existence on anything else. And in that sense he's ontologically ultimate, the ultimate explanation of all other things."

I asked Swinburne to consider, just for the sake of argument, another possibility: that the universe exists as a brute fact, without any God to sustain it. Would the universe itself then be necessary in his sense, since it would not depend on anything else for its existence?

"That's right!" he replied.

So the God hypothesis—even if it is accepted as more probable than the alternative, that of a complex universe which exists uncaused—does not fully resolve the mystery of existence.

"I must admit," Swinburne said, "that part of me wants to know, wants some guarantee that there couldn't *not* be a God. But I understand that it's not logically possible to explain everything. You can explain A by B, B by C, and C by D, but in the end all you can do is find the simplest hypothesis that explains as much as possible of reality. That's where explanation *has* to stop. And that intellectual stopping point, I claim, is God. As to why God exists, I can't answer that question. I can't answer that question."

Could even Swinburne's God, if we were able to ask him, answer it? "I am who am," the voice from the burning bush announced to Moses. But did that voice ever ask, "Whence then am I?" If there were an explanation for God's existence, then God, being omniscient, would know it. But if there really was no explanation—if he is indeed the Supreme Brute Fact—then he would know that too. He would know that his own existence as a contingent being was, in Swinburne's words, "vastly improbable." Would the divine mind be puzzled by its inexplicable triumph over the perfect simplicity of Nothingness?

I didn't pursue this potentially impious line of questioning. I had made enough demands on Swinburne's hospitality, on his supply of tea and biscuits, and perhaps on his intellectual patience as well. The windows of his study had grown dark with the early sunset. It was time to go. I thanked him effusively, and he gave me some advice on which restaurants I might try that evening in Oxford.

The birdsong had long since quieted down when I left Swinburne's apartment building. Ambling back onto the main road, I noticed again the prominent Eastern Orthodox Church that loomed nearby. It seemed an odd intrusion of Byzantium into North Oxford. Swinburne had told me he was an Orthodox communicant. Did he worship there? With his sacerdotal manner and his elongated, slightly severe features, this Oxford philosopher of science and religion could almost take his place in an Eastern church mosaic, right next to the other Byzantine divines:

O sages standing in God's holy fire
As in the gold mosaic of a wall . . .

Say, was that "a great cathedral gong" I heard in the distance?

No, it was just the bells of Oxford summoning me back toward the High Street. On reaching this destination, I went into one of the restaurants Swinburne had recommended, the Quod Brasserie. It was half full and fairly lively, provincial in an academic way that contrasted with the cosmopolitan Café de Flore back in Paris. Taking a table by myself, I ordered smoked haddock and a tomato salad, along with a split of champagne and a full bottle of Australian Shiraz, and mindlessly read that day's issue of the *Guardian* as I ate and drank. By the time I left the place, it was close to midnight. Walking down the almost deserted High Street back toward my hotel, I was engulfed by a diffuse sense of contentment, and I temporarily ceased to care about the mystery of existence.

Interlude

The Supreme Brute Fact

Richard Swinburne seems to have solved one mystery at the price of introducing another. He purports to explain the world's existence by positing a God who created it. But he concedes that he can find no explanation for God himself, whose existence, compared to the stark simplicity of Nothingness, strikes Swinburne as "vastly improbable." Is this the best that theism can do—cap off its cosmic explanation with an inexplicable being, a Supreme Brute Fact?

Traditional philosophers of theism have not thought so. They have held that God, unlike the world, exists by his very nature. He contains within himself the principle of his own being. There are many technical terms for this. God is *causa sui*, the "cause of himself." He possesses *aseity*, the property of being self-existent. He is the *ens realissimum*, the most real being, and the *ens necessarium*, the necessary being.

But is there any justification for all this verbiage?

Consider, for example, the term *causa sui*. It seems to suggest that God somehow caused himself to exist. But even medieval theologians refused to go that far. No being, they held, could possibly bootstrap itself into existence. Regardless of how powerful the being in question might be, it would have to exist before it could exercise its causal powers.

To say that God is *causa sui* is really to say that he is uncaused. His existence needs no cause because it is necessary. Or, to put the point somewhat differently, his existence needs no explanation because it is self-explanatory.

And how might the existence of such a self-explanatory being be demonstrated? One traditional route is the *cosmological argument* for the existence of God. This argument goes back to Aristotle, but the most sophisticated version of it is due to Leibniz, and it goes like this.

The universe is contingent. It might not have existed. Given that it does exist, there must be an explanation for its existence. It must have been caused to exist by some other being. Suppose that this being, too, is contingent. Then it requires an explanation for its existence as well. And so on. Now, either the explanatory chain eventually comes to an end, or it does not. If it does come to an end, the last being in the chain must be self-explanatory. If it goes on *ad infinitum*, then the entire chain of beings stands in need of an explanation. It must have been caused by some being outside the chain. Then the existence of *that* being must be self-explanatory. In either case, the existence of a contingent world must ultimately be explained by something whose existence is self-explanatory.

Once the existence of a self-explanatory being is deduced, just a little logical tinkering is needed to show that this being has the properties traditionally ascribed to God. (It was Samuel Clarke, an English theologian and friend of Isaac Newton, who supplied the details.) Start by observing that a self-explanatory being must exist as a matter of necessity. And if it exists necessarily, it must exist always and everywhere—that is, it must be *eternal* and *infinite*. It must also be *powerful*, since it caused the contingent world to come into existence. Moreover, it must be *intelligent*, since intelligence exists in the world and therefore must exist in its cause. And since it is also infinite, it must be *infinitely powerful* and *infinitely intelligent*. Finally, it must be *morally perfect*. For, being infinitely intelligent, it can never fail to apprehend the truth as to what is good; and, being infinitely powerful, it can never be prevented by any weakness of its own from acting in accordance with that truth.

The preceding reasoning, intended to show that the necessary being deduced in the cosmological argument must be God-like, is obviously rife with fallacy. But how about the cosmological argument itself? How valid is it? In essence, Leibniz was attempting an inference from contingency to necessity: *if* there is a contingent world, and *if* everything has an explanation, *then* there must be a necessary being that explains the existence of this world. Leibniz's first premise looks okay. There *does* seem to be a world, and it *does* seem contingent. The second premise, which is Leibniz's famous Principle of Sufficient Reason, is more dubious. Even Swinburne denied that there was an explanation for absolutely everything. And without that premise, the cosmological argument collapses.

But, valid or not, there is something peculiar about the cosmological argument. It is supposed to take us from an empirical premise—arising from our experience of the actual universe—to a necessary being. But if there *is* such a necessary being, why do we need this empirical premise to deduce its existence? Why can't we infer its existence directly, through pure reason?

There is, as it happens, a notorious bit of reasoning which attempts to do just that. It is called the *ontological argument*. Unlike the cosmological argument for the existence of God, the ontological argument has no need of the premise that a world exists, or of the premise that there is an explanation for everything. The ontological argument purports to establish God's existence through logic alone. God must exist as a matter of logical necessity, it says, since he possesses all perfections, and it is more perfect to exist than not to exist.

The ontological argument was invented in the eleventh century by Saint Anselm, an Italian monk who eventually became the archbishop of Canterbury. The gist of it seems to have come to this monk one day during his morning prayers. God, reasoned Anselm, is by definition the greatest and most perfect thing that can be conceived. Now, suppose God were merely an object of thought—something, that is, which existed only in our imagination. Then it would be possible to conceive of another being exactly like God except that this being existed in reality too. And since it is greater to exist in reality than merely in the imagination, this being would be greater than God—which is absurd. Therefore, God's nonexistence is a logical impossibility. "So truly, therefore, dost thou exist, O Lord, My God, that thou canst not be conceived not to exist," concluded the prayer in which Anselm expressed his argument.

Could the ontological argument possibly be valid? Even those who believe in God may feel it is too good to be true. Aquinas did not accept it. Descartes did, although he put it into a somewhat different form. Leibniz felt it needed an extra premise, namely, that God is a *possible* being—which premise Leibniz easily supplied by showing that God's various perfections were all compatible with one another. Schopenhauer dismissed the ontological argument as "a charming joke." Bertrand Russell, by contrast, describes in his autobiography how as a young man he was struck by its seeming truth:

> I remember the precise moment, one day in 1894, as I was walking along Trinity Lane, when I saw in a flash (or thought I saw) that the ontological argument is valid. I had gone out to buy a tin of tobacco; on my way back, I suddenly threw it up in the air, and exclaimed as I caught it: "Great Scott, the ontological argument is sound."

Later in his philosophical career, Russell decided that the ontological argument was not sound after all. Still, he observed, "it is easier to feel convinced that it must be fallacious than it is to find out precisely where the fallacy lies."

Russell's observation has been borne out by contemporary anti-theists, whose critique of the ontological argument often boils down to mere mockery. For instance, Richard Dawkins, in *The God Delusion*, dismissed the ontological argument as "infantile," a bit of "logomachist trickery," but he did not take the trouble to identify the defect in its logic. The very idea that "a grand truth about the cosmos should follow from a mere word game" struck Dawkins as simply ridiculous, and that was the end of the matter as far as he was concerned.

But what exactly *is* wrong with the ontological argument? Anselm's reasoning, succinctly put, runs like this:

1. God is the greatest imaginable being.

2. A being that exists is greater than one that is merely imaginary.

Therefore:

3. God exists.

Premise (1) can scarcely be disputed, since it embodies the very definition of God. Premise (2), though, looks a little funny. Just how much greater is it to exist in reality than to exist merely in the imagination? Am I, by dint of my reality, greater than the imaginary Emperor of Ice Cream?

Think for a moment about the phrase "exists merely in the imagination." While it's a familiar enough locution, it has distinctly odd implications if taken literally. It suggests that the being in question is real, yet somehow confined to a tiny piece of territory—our heads. And clearly such a cerebrally confined being is less great than one that is free to manifest itself in the cosmos at large. But that can't be right. What is in our heads is not the thing itself, but the *idea* of the thing. And the idea is nothing like the thing. (You can ride a unicorn, for example, but you cannot ride the *idea* of a unicorn.) To say that a being "exists merely in the imagination" is really a *façon de parler*. It does not entail that the being in question exists in some limited way. Rather, it asserts that we have a certain idea/concept/image in our minds, but that no being corresponds to this idea/concept/image. An idea of God is not a kind of God, albeit less perfect, any more than a painting of a piece of fruit is a kind of fruit, albeit less nutritious.

Suppose, however, we forget about "imaginary existence" and simply concede that it is more perfect to exist than not to exist. Then God, possessing all perfections, must exist, mustn't he? So what is wrong with Anselm's reasoning?

The most celebrated objection to the ontological argument was brought by Kant. Existence, Kant claimed, is not a real predicate. In other words, *being existent* is not an ordinary property of things, like *being red* or *being intelligent*. This objection is routinely cited by all who would dismiss the ontological argument—Dawkins, for example. If existence is not a property of any sort, then it can't very well be a perfection.

Is Kant's dictum—existence is not a predicate—valid? Existence certainly seems a peculiar property in one sense: it is universal. Unlike the properties of *redness* or *intelligence*, absolutely everything has it. Just try to name something that does not exist. Santa Claus? To say "Santa Claus does not exist" is not to attribute nonexistence to some entity; it is merely to say that nothing satisfies the description *jolly fat man who lives with elves at the North Pole and who distributes toys to children the world over on Christmas Eve*. Even to say, "There is something that does not exist," is self-contradictory, since the "there is" part asserts the very existence that the "does not exist" part denies.

It is not obvious why the mere fact that *existence* is universally possessed should disqualify it from the honor of being a property. But Kant evidently had something different in mind when he said "existence is not a real predicate." His point seemed to be that existence adds nothing to the content of a concept. "A hundred real dollars do not contain the least coin more than a hundred possible dollars," he wrote, adding, "My financial position is, however, affected very differently by a hundred real dollars than it is by the mere concept of them."

And here Kant is certainly right. Suppose I take a concept like *current member of the United States Senate*. There are precisely one hundred individuals of whom this concept is true. Now suppose I add *existence* to the concept, getting *current existing members of the United States Senate*. Lo and behold, this new concept is true of the same one hundred individuals that the old one was!

So adding *existence* to a concept doesn't give it any extra heft. Nor does it fortify the existential chances of the would-be object defined. Otherwise, we could bring into being all sorts of wonderful things merely by defining them in the right way. This point was made by Saint Anselm's earliest critic, a fellow eleventh-century monk called Gaunilo of Marmoutier. By Anselm's logic, Gaunilo observed, we could demonstrate that somewhere on the ocean there must be an ideally pleasant "lost island," since actual existence necessarily numbers among this island's perfections.

What happens, from a logical point of view, when we deny God's existence? Well, suppose we define God in the same theologically orthodox way that Saint Anselm did, as an infinitely perfect being. And, just to give Anselm's side the advantage, let's explicitly build *existence* into his definition:

x is God if and only if x is infinitely perfect and x exists.

Then to say, "There is no God," is to say:

There is no x such that x is infinitely perfect and x exists.

But this is equivalent to:

For every x, either x is not infinitely perfect or x does not exist.

And there is nothing inherently self-contradictory about that proposition. Indeed, it would be true of a world in which every entity fell short of infinite perfection—which is precisely the sort of world that atheists claim we live in.

Still, there is a reason why Anselm felt it was self-contradictory to deny the existence of God. That is because we use "God" not just as shorthand for a description—*infinitely perfect being*—but also as a name. If God is infinitely perfect and therefore existent, how could he *fail* to exist?

To see what's wrong with this way of thinking, consider a description that is formally similar: *oldest living man*. Suppose we decide to call the oldest living man (whoever he might be) "Methuselah." Now ask the question, Is Methuselah alive? Well, of course he is. By definition, he's the oldest *living* man. How could he fail to be alive? But if Methuselah cannot fail to be alive, then he can't possibly be dead. He must be immortal! Such are the logical perils of sticking a name on a definition.

So the ontological argument, in its classic Anselmian version, is unsuccessful. Even if existence is built into the very definition of God, it does not follow that there is a being that satisfies this definition. Is that the end of the matter?

As it happens, no. In recent decades, the ontological argument has been resurrected in an apparently more powerful form. The new version relies on a kind of logic undreamt of by Saint Anselm: *modal* logic. Modal logic outstrips the resources of ordinary logic. Whereas ordinary logic concerns itself with what is and is not the case, modal logic deals with what *must* be the case, what *might* be the case, and what *could not possibly* be the case—a far stronger set of notions.

Modal logic was developed by some of the greatest twentieth-century logicians, including Kurt Gödel and Saul Kripke. It was Gödel, the author of the notorious "incompleteness theorems," who saw in modal logic a way of reviving the ontological argument in a strengthened form. The idea seems to have come to him in the early 1940s, but he did not divulge it until a few years before his death (by self-starvation) in 1978. Whether Gödel was convinced by his own version of the ontological argument is unclear. But he was certainly open to the existence of God, maintaining that it might be possible "purely rationally" to reconcile the theistic worldview "with all known facts."

Gödel is not the only one to have noticed the theological uses of modal logic. Independently of him, several philosophers have come up with similar modalized updates of Anselm's reasoning. The most prominent of them is Alvin Plantinga, a professor at the University of Notre Dame. Plantinga's efforts to secure the existence of God by logic alone have even attracted the attention of *Time* magazine, which hailed his "tough-minded intellectualism" and called him "the leading philosopher of God."

The modal ontological argument for God's existence can look dauntingly technical. Gödel expressed the argument in a series of formal axioms and theorems, and Plantinga took the better part of his treatise *The Nature of Necessity* to lay out all the details. Still, the nub of it can be put in a fairly simple form.

A truly great being, the argument begins, is one whose greatness is robust in the face of chance. Such a being not only *is* great, but it *would have been* great even if events had turned out differently from the way they actually did. By this criterion, for example, Napoléon was not truly great, since he might have died of the flu as a child in Corsica instead of growing up to conquer Europe. Indeed, if his parents had arranged their schedule of sexual congress differently, Napoléon might not have existed at all.

Now, a *maximally great* being is one whose greatness is unexcelled in *every possible* world. Such a being would, if it existed, be omniscient, omnipotent, and perfectly good. And there would be no possible state of affairs in which these maximal qualities were in any way diminished. It follows that such a being could not be merely contingent, existing (like Napoléon) in some possible worlds and not others. If such a maximally great being existed at all, it would exist *necessarily*, in every possible world.

For brevity's sake, let us call such a maximally great being "God." So far, so good. Now comes the twist. Does God exist? "Almost certainly not," an atheist like Richard Dawkins would say. But even Dawkins concedes that, improbable as God's existence might be, it is at least *possible* that there is a God—just as it is possible (but highly unlikely) that there is a celestial teapot in orbit around the sun.

But this is a fatal concession for the atheist to make. To say it is possible that a celestial teapot is orbiting the sun is to say that in some possible world such a teapot *is* orbiting the sun. And to say it is possible that God exists is to say that in some possible world there *is* a God. But God is different from a teapot. He is by definition a maximally great being. Unlike a teapot, his greatness—and therefore his existence—is stable across different possibilities. So if God exists in *some* possible world, he must exist in *every* possible world—including the *actual* world. In other words, if it is even possible that God exists, then it is *necessary* that he exists.

That is the rather breathtaking conclusion of the modal ontological argument. And it is an entirely valid one, at least within the framework of modal logic. (To be specific, it is valid in the system of modal logic known in the trade as "S5.") As Plantinga correctly observes, "It breaches no laws of logic, commits no confusions and is entirely immune to Kant's criticism."

Unlike Anselm's ontological argument, the modal version does not take *existence* to be a predicate or a perfection. It does take *necessary existence* to be a perfection, but that is entirely plausible. Whereas *existence* is not a great-making quality—everything has it, after all—*necessary existence* obviously is great-making. To exist necessarily means that your existence depends on nothing else. It could not have been prevented. You are immune from the possibility of annihilation. Finally, and not least among its virtues, the modal ontological argument holds out the hope of answering the question *Why is there something rather than nothing?* If God is possible, it says, then God is necessary—and hence nothingness is impossible.

Is God possible? Or—to put it in the jargon of the modal ontological argument—is maximal greatness possibly exemplified? Think a bit about what "maximal greatness" means. A maximally great being is one that, if it exists in any possible reality, exists in all of them. It's analogous to a being that, if it can be found anywhere in the world, manages to be everywhere, including here; or to a being that, if it exists at any moment in history, must exist at all moments, including the present one. A maximally great monarch would be one who, if he had a kingdom anywhere in the universe, would reign over the entire universe. A maximally great man, if he ever lived, would live eternally.

Clearly, maximal greatness is well beyond the realm of the familiar. How then could we know that such a thing is possible? Gödel concocted an elaborate argument to prove that the idea of a maximally great being was not inherently self-contradictory (the way, say, the idea of the largest number is inherently self-contradictory). Hence, Gödel concluded, such a being is logically possible. And since the range of possible worlds covers every logical possibility, there is a world that contains a maximally great being. But if such a being exists in any possible world, it must exist in every possible world—including our own, the actual world.

Unhappily for partisans of the ontological argument, this logic cuts both ways. There is nothing inherently self-contradictory either in the supposition that a maximally great being does *not* exist. Indeed, Plantinga himself refers to the property of there *not* being a maximally great being by the term "no-maximality." So, by parity of reasoning, there must be a possible world in which no-maximality is exemplified—that is, one in which maximal greatness is absent. But if God is absent from *any* possible world, he is absent from *all* possible worlds—in particular, he is absent from the actual world.

So which will it be? If, in the framework of modal logic, we accept the premise that God possibly exists, then we are committed to the necessity of his existence. If we accept the premise that God possibly does not exist, then we are committed to the impossibility of his existence. Both can't be true. Yet, from a purely logical perspective, the possibility of God's existence seems no more compelling than the possibility of his nonexistence. Should we simply flip a coin to decide which premise to accept?

Recognizing the force of the counterargument, Plantinga has conceded that "a sane and rational man" might well reject the premise that a maximally great God is possible, and that the "canny atheist" will certainly do so. Without that premise, of course, this contemporary version of the ontological argument collapses. Nevertheless, Plantinga advocates accepting the premise in the interests of "simplifying" theology—the way one might accept a wild-sounding premise of quantum theory in the interests of simplifying physics.

Critics of the modal ontological argument will have none of this. "The premise that it is just *possible* that there should be something unsurpassably great looks innocent," observed the Oxford philosopher (and staunch atheist) John Mackie. But this premise, he warns, is a Trojan horse: "Anyone who is not already and independently persuaded that traditional theism is true has good reason to reject the key premise" of the modal ontological argument. Thus, although the argument may be "interesting . . . as a logical peculiarity," Mackie said, it is "worthless as a support for theism."

There is a deeper issue lurking here. Could logic alone answer the question *Why is there something rather than nothing?* Can pure thought secure the existence of a positive reality that necessarily prevails over nothingness? "Every philosopher would *like* to say yes," Bertrand Russell observed, "because a philosopher's job is to find out things about the world by thinking rather than observing." If "yes" is the right answer, Russell added, then there is a "bridge" from pure thought to concrete existence.

How sturdy is the bridge offered by the ontological argument? The God it purports to deliver is a necessary being. His existence is a truth of pure logic, a tautology. But tautologies are empty propositions. Since they are true regardless of how reality is, they are devoid of explanatory content.

How could such a tautological divinity be the *fons et origo* of the contingent world we see around us? How could a tautology exercise free will in creating it? The gap between necessity and contingency is no less difficult to bridge than the gap between being and nothingness.

The God of Richard Swinburne is most decidedly unlike the God of the ontological argument. Swinburne's God is not the product of logic. He has a free will that transcends any tautology. He exists in time. He is not even maximally great, at least in the sense demanded by the ontological argument, since his omniscience is limited by his inability to know in advance how we, his creatures, will exercise our own free will. He is a fitting ontological foundation for a contingent world. Yet he himself has no ontological foundation. His essence does not include existence. His being is not logically necessary. He might not have existed. There might have been no God, nothing at all.

Swinburne posits such a God because, he claims, that is the "simplest stopping point" in the task of explaining the existence of the world and the way it is. The God hypothesis is the one that minimizes the part of reality that is left unexplained. But by positing God, Swinburne has added a new and unexplained element to the picture. Kant was right: the cosmological argument for God's existence works only when it is backed up by the ontological argument. If the ontological argument fails, God is not a necessary, and therefore self-explanatory, being. Then the child's seemingly naive question—"But Mummy, who made God?"—remains a live one. Which raises a tantalizing thought: could there be some deeper explanatory factor that encompasses both the world and—if he really exists—God too? How deep can explanation go?

There was another man in the Oxford vicinity who, I had heard, was qualified to answer that question. But, before I could talk to him, it seemed that I myself had some explaining to do.

7

THE MAGUS OF THE MULTIVERSE

What if there is no limit to what can be explained? What if reality turned out to be comprehensible through and through? Indeed, what if reality were to *mandate* its own comprehensibility?

Sheer fantasy, you might say, an epistemic pipe dream. Only a fool could believe that reality can be made to yield up all its secrets to creatures like us living within it.

Yet I knew there was someone in the vicinity of Oxford who *did* believe this, someone who was far from being a fool. His name is David Deutsch, and he is widely regarded as one of the most daring and versatile scientific thinkers alive. "Deutsch seems more passionate about what reality is, about what actually exists and why, than almost any other scientist I can remember meeting," one veteran journalist wrote of him. And Deutsch is also a man with a singular achievement to his credit: in 1985, he demonstrated the theoretical existence of a *universal quantum computer*—a computer capable of simulating any physically possible reality.

The idea of a computer that would harness the weird power of quantum mechanics was not original to Deutsch. It was Richard Feynman who, around the beginning of the 1980s, first seems to have dreamed it up. At the time, Deutsch had recently graduated from Cambridge University. Having barely scraped a "pass" degree in mathematics, he traveled to the United

States, where he sought out distinguished physicists like John Archibald Wheeler and Bryce DeWitt.

While studying how quantum fields behave in curved spacetime, Deutsch became obsessed with the "many worlds" interpretation of quantum theory. This interpretation was the 1950s brainchild of Hugh Everett III, a Princeton graduate student who went on to become a strategic planner at the Pentagon before dying in 1982. According to the many-worlds interpretation, our universe is merely one among a vast ensemble of alternate universes—a *multiverse*—all of them interacting in a ghostly way to produce otherwise inexplicable quantum phenomena.

What would happen, Deutsch wondered, if quantum mechanics was applied to computer science? Might all the different parallel universes in the multiverse be coaxed into collaborating on a single computation?

Deutsch took as his starting point the classical theory of computability, which had been pioneered in the years before the Second World War by the Englishman Alan Turing. Among Turing's discoveries was a program for a "universal" computer, one that would be capable of mimicking to perfection the output of any special-purpose machine. Deutsch set about recasting Turing's work in quantum terms. In doing so, he managed to construct a quantum version of Turing's universal computer—that is, a single quantum operator (or "Hamiltonian," as it is known in the trade) that can do the work of any conceivable computing machine, whether a conventional computer of the kind now in use or a quantum computer as envisaged by Feynman. And Deutsch's universal quantum computer had another marvelous property: in principle, it could simulate any physically possible environment. It was the ultimate "virtual reality" machine.

Deutsch, who was in his early twenties at the time (he was born in Israel in 1953), later downplayed his proof of the existence of a universal quantum computer as "fairly straightforward." He went to Caltech to present it to Richard Feynman, who was already suffering from the cancer that would kill him in 1988. After Deutsch had written the first bits of his proof on the blackboard, the ailing Feynman startled him by jumping out of his seat, grabbing the chalk, and finishing it himself.

For Deutsch, a universal computer had become nothing less than the key to understanding reality. Such a machine, being able to generate all physically possible worlds, would be the consummation of physical knowledge. It would be a single, buildable physical object that could describe or mimic with perfect accuracy any part of the quantum multiverse. And since it was *possible* to build a universal computer, Deutsch concluded, such a machine must actually *be* built somewhere in the multiverse. Omniscience exists!

Such speculative flights come quite naturally to Deutsch, who, after returning to England from the United States, was appointed to be a research physicist at Oxford's Clarendon Laboratory. In 1997, he laid out his worldview in a book titled *The Fabric of Reality*. To achieve a deep scientific understanding of reality, he argued therein, we must use not only quantum mechanics and the theory of computation, but also the theory of evolution. (He credits Richard Dawkins as one of his intellectual heroes.) Life and thought, he declared, determine the very warp and woof of the quantum multiverse. Whereas physical structures, like constellations and galaxy clusters, vary randomly from one universe to the next, knowledge-bearing structures—embodied in physical minds—arise from evolutionary processes that ensure they are nearly *identical* across different universes. From the perspective of the quantum multiverse as a whole, mind is a pervasive ordering principle, like a giant crystal.

Clearly, here was a man who aspired to a complete understanding of what he was pleased to call the "fabric of reality." Would that complete understanding encompass the mystery of existence itself? Would it yield an answer to the question *Why is there something rather than nothing?* I ardently hoped to find out. I had reviewed Deutsch's book years ago in the *Wall Street Journal*—favorably, as I dimly recalled. Surely, I thought, he would be willing to talk to an admirer such as myself, especially one who had taken the trouble to come all the way to Oxford. So I e-mailed him, introducing myself and mentioning the nice review I had given his book in the United States more than a decade ago.

"I just checked on Google," Deutsch e-mailed me back. "*Arrogant in tone and marred by leaps of logic*—is that the one?"

Oh dear. My memory seemed to have played me false. I googled the review myself. The full sentence he had quoted read, "Arrogant in tone and marred by leaps of logic, his book nonetheless bristles with subversive insights about virtual reality, time and time travel, mathematical certainty, and free will." That didn't sound so bad. In the review I had also called Deutsch "mad, bad, and dangerous to know"—a description originally applied to Lord Byron. E-mailing him again, I pointed out that this was meant, in a somewhat jocular vein, as a compliment.

"In my opinion Byron was *literally* mad, bad, and dangerous to know, not least because he was a willfully careless thinker," Deutsch replied in a second e-mail. "So I don't consider being compared to Byron to be any sort of compliment."

This wasn't going well. But when tact and flattery fail, I have found, abject groveling sometimes succeeds. So, gushing apology, I simply implored him to meet me.

"No problem, I'd be interested to have a chat," he replied. "But I'd like something in return. Please let me know specifically what the first leap of logic in *The Fabric of Reality* is, and the place where it first becomes clear to you that its tone is arrogant."

Fortunately, I had brought along to Oxford my old reviewer's galley of the book. Holed up in my minuscule hotel room on the High Street near Logic Lane, I spent a stressful afternoon trying to decipher the critical comments I had long ago illegibly scribbled in the galley's margins. Finally I found what seemed to me a "leap of logic." Deutsch's "Turing principle" implied that there is no limit to the number of computational steps that are physically possible. And that, in turn, implied that universe must eventually collapse upon itself in a Big Crunch, since only such a fiery ending could furnish the infinite energy needed for infinite computation. So, Deutsch concluded, such a Big Crunch *must* be our cosmic fate. But that can't be right, I thought. Current cosmological evidence points to a contrary fate for our universe: rather than eventually collapsing upon itself, it will expand forever, dissipating into a chilly void. If Deutsch's logic entailed the opposite conclusion, surely there must have been an unjustified leap in it somewhere.

I e-mailed Deutsch to this effect. He conceded that there might be something to my criticism, while observing that it applied to a claim he had made rather late in the book. "Could it be that the first leap of logic was in the last chapter?" he asked.

Nevertheless, he was gracious enough to invite me to his house for tea. And, after fleetingly entertaining the paranoid suspicion that he might be out to poison me—an author's fit revenge against an impertinent reviewer— I accepted.

It turned out that Deutsch did not actually live in Oxford, but in a nearby village called Headington, where, an Oxford friend informed me, J. R. R. Tolkien and Isaiah Berlin had made their homes. I decided to go there on foot. Crossing the Magdalen Bridge over the Cherwell, I paused for a moment to watch some students floating lazily down the river in their punts. Then I rounded a traffic circle at the city limits and followed the curving road up a hill, making my way along an ancient-looking stone wall. A woman cyclist passed me on the road, with a log and some tree branches strapped to her bike, which reminded me of the "log lady" in *Twin Peaks*. After continuing on a few miles, I attained a sort of plateau, where I came upon a collection of little brick houses, a restaurant called Café Bonjour, and a Domino's Pizza parlor. This was Headington.

Upon reaching the address Deutsch had given me, I found a small two-story house hidden behind some shaggy trees. A trio of flags hung from the front of the house—British, Israeli, and American. A junked TV set sat outside. I tried the doorbell, but it didn't work. So I rapped on the dimple-frosted glass.

After a few moments, the door was opened by an improbably boyish-looking fellow with large mole-like eyes, rather transparent skin, and shoulder-length, albinoid hair. Behind him, I could see great moldering heaps of papers, broken tennis rackets, and other detritus. I knew that Deutsch was famous for, as one science journalist put it, "setting international standards in slovenliness," but these looked more like experiments in indoor composting.

He beckoned me inside and led me past the piles of rubbish into a room with a large television and an exercise bike. On a sofa sat an attractive young woman with strawberry-blonde hair—she looked almost like a teenager—eating a plate of macaroni and cheese. Deutsch addressed her as "Lulie." She moved over to make room on the sofa for me, and the conversation began, albeit on a discouraging note.

"On the question of why there is something rather than nothing, I'm not sure I know anything apart from that joke," Deutsch opened. "How does it go? Oh yeah—'Even if there was nothing, you'd still be complaining!'"

I told him the joke came from Sidney Morgenbesser, an American philosopher who had died a few years ago.

"Haven't heard of him," Deutsch said.

But how could Deutsch be so cavalier about the mystery of existence? After all, he didn't believe that there was just *one* world. His view of reality encompassed a *huge ensemble* of worlds, all existing in parallel: a multiverse. The multiverse was for Deutsch what God had been for Swinburne: it was the simplest hypothesis that explained what we observed around us—notably, the weird phenomena of quantum mechanics. If the physical laws governing the multiverse mandated their own comprehensibility, as Deutsch believed, shouldn't they also mandate the comprehensibility of reality as a whole?

"I don't think that an ultimate explanation of reality is possible," he said, shaking his head. "That doesn't mean I think there's a *limit* to what we can explain. We'll never run into a brick wall which says, 'NO EXPLANATION BEYOND THIS POINT.' On the other hand, I don't think we'll find a brick wall that says, 'THIS IS THE ULTIMATE EXPLANATION FOR EVERYTHING.' In fact, those two brick walls would be almost the same. If, *qua impossibile*, you were to have an ultimate explanation, it would mean the philosophical problem of why *that* was the true explanation—why reality was this way and not another—would be forever insoluble. Hello, I hear the water boiling!"

He went into the kitchen. Lulie smiled at me and continued to pick at her macaroni.

When Deutsch emerged a few moments later with a teapot and a plate of biscuits, I asked him whether he was puzzled at all by the existence of the multiverse. Was the question *Why is there something rather than nothing?* a profound one, or was it simply misguided?

"Hmmm," he responded, touching his temple, "... a deep question ... a misguided question. . . . Look, I can't rule out the possibility that there is a foundation for reality. But if there is, the problem of *why* that's the foundation would still be insoluble."

He took a sip of tea and continued, "Take the 'first cause' argument, the idea that the existence of the world must be explainable by some sort of originating event. It's hopelessly parochial! The idea that things are always caused by things that come before them in time has nothing to do with logic or explanation as such. You could imagine an explanation where something was caused by things happening at all different times, past and future. Or an explanation that didn't have anything to do with time at all, or even with causes. The real question you want to answer is not what came *before*, but why something *is the way it is*."

I gingerly sipped at my cup of tea, which did not seem to be poisoned.

"You can't give a once-and-for-all definition of what an explanation is," Deutsch said. "In fact, important explanatory advances often change the *meaning* of explanation. My favorite example is the Newtonian-Galilean revolution, which not only brought in new laws of physics, but also altered the very notion of what a physical law is. Previously, laws had been rules stating what happens. Kepler's laws, for instance, were about how the planets traveled around the sun in elliptical orbits. Newton's laws were different. They didn't talk about planets or ellipses. Instead, Newton's laws were rules that *any* such system would obey. It's a different style of explanation, one that hadn't been thought of before, one that wouldn't even have been considered an explanation before. The same kind of explanatory revolution happened a couple of hundred years later with Darwin. Previously, when people asked, 'Why does this animal have the shape it does?' they expected that the answer would cite some *property* of the shape—that it was efficient, that it was favored by God, and so forth. After Darwin, the answer wasn't about properties of the shape, but about how that shape had come into existence by evolution. Again, it's a different style of explanation."

Deutsch paced back and forth as he spoke. I remained seated on the sofa next to Lulie, who had finished her plate of macaroni and cheese.

"This point about the fluid nature of explanation is a real hobby horse of mine," he continued, his voice gaining in intensity. "I think we're going to need a different style of explanation to solve problems like free will and consciousness. These are fundamentally philosophical problems, not technical problems. I don't think artificial intelligence will be achieved until philosophical progress is made in understanding what consciousness *is*. We couldn't make artificial life without the concept of a replicator, and we don't have the equivalent concept yet for consciousness. You can't program what you can't specify."

This struck me as refreshingly at odds with the prevailing orthodoxy of the artificial-intelligence community, whose members seemed to think that the mystery of consciousness would wither away with the advent of superintelligent computers, which was supposedly just around the corner.

But back to the multiverse. Where did it come from? Why is there a "fabric of reality" at all?

"To my way of thinking," Deutsch said, "that question could only be answered by finding a more encompassing fabric of which the physical multiverse was a part. But there is no ultimate answer."

Could he then see what form that larger fabric of reality might take?

"I would start with the principle of comprehensibility," he said. "Look, there's a quasar out there in space, billions of light-years away. And in our brain there's a model of the quasar, a model that has remarkable properties. There's not just an image of the quasar in our brain, there's a structural model with the same causal and mathematical relationships. So here you have two objects that are physically as dissimilar as they could possibly be—a quasar, which is this black hole with jets, and our brain, which is chemical scum—and yet they embody the same mathematical relationships!"

Interesting point, I interjected, but the relevance of it escaped me.

"For that to happen, the laws of physics must have a very special property. They permit—they *mandate*—their own comprehensibility. And you can take this further. If it's true that the world is comprehensible, that we're capable of understanding it, then in order to understand the behavior of humans, you need to understand everything! Since the structure of quasars is represented in the brains of human scientists, the behavior of scientists depends on the behavior of quasars. To predict what papers a physicist will write next year, you have to know something about quasars.

By the same argument, it follows that to know all truths about humans, you need to know all the truths there are."

Deutsch paused, as if to regather his thoughts. "We're bootstrapping our way toward better and better explanations. And that's why we can never have an ultimate explanation. Anything pretending to be an 'ultimate' explanation would be a *bad* explanation, because there would be nothing left over to explain why it was the right one—to explain why reality was that way and not another way."

Deutsch had long maintained that quantum theory was a key to understanding the fabric of reality. And in quantum theory, I observed, you can seemingly get Something from Nothing. A particle and its antiparticle, for instance, can spontaneously appear out of the vacuum. Some physicists have conjectured that the universe itself began as a vacuum fluctuation— that it "tunneled" into existence out of nothingness. Might quantum theory explain why there was a world at all?

"Not the least!" he replied. "Quantum theory is too parochial to address the question of existence. When you talk about a particle and an antiparticle appearing in the vacuum, that's not at all like coming into existence out of nothing. The quantum vacuum is a highly structured thing that obeys deep and complex laws of physics. It's not 'nothingness' in the philosophical sense at all. It's not even as little as the kind of nothing you have in your bank account when there's no money in it. I mean, there's still the bank account! A quantum vacuum is much more even than an empty bank account, because it's got *structure*. There's stuff *happening* in it."

So the laws governing the quantum multiverse can't tell us anything at all about why the multiverse exists?

"No, none of our laws of physics can possibly answer the question of why the multiverse is there," he said. "Laws don't do that kind of work." He recalled an image from the great John Archibald Wheeler, his onetime mentor. "Wheeler used to say, take all the best laws of physics and write them down on bits of paper and put those bits of paper on the floor. Then stand back and look at them and say, 'Fly!' They won't fly. They just sit there. Quantum theory may explain why the Big Bang happened, but it can't answer the question you're interested in, the question of existence.

The very concept of existence is a complex one that needs to be unpacked. And the question *Why is there something rather than nothing?* is a layered one, I expect. Even if you succeeded in answering it at some level, you'd still have the next level to worry about."

Click! My tape recorder switched off. Somewhat depressingly, it had reached the end of side B of the microcassette without registering a single genuine advance toward a resolution to the mystery of existence.

Should I have been surprised? In the opening pages of *The Fabric of Reality*, after all, Deutsch had written, "I do not believe that we are now, or ever shall be, close to understanding *everything there is.*" Still, he had managed to impress one positive lesson on me: that there is a lot more to reality than we might imagine. The part of it that we inhabit not only is tiny, but also may be grossly unrepresentative of the whole, giving us a partial and distorted view. We are like the prisoners chained inside the cave of illusion in Plato's famous allegory. It may even be—although Deutsch told me that he thought it improbable—that we exist within a *simulated* reality, one created by higher beings—beings who, like Descartes's evil genie, have deliberately programmed it with the wrong laws of physics. Yet even if we were inmates confined to such a partial and distorted reality, our quest for understanding would eventually take us beyond its virtual walls.

"It is not enough that the inmates be prevented from observing the outside," he had written in *The Fabric of Reality*. "The rendered environment would also have to be such that no explanations of anything inside would ever require one to postulate an outside. The environment, in other words, would have to be self-contained as regards explanations. But I doubt that any part of reality, short of the whole thing, has that property."

But if the whole of reality *were* explanatorily self-contained, then it would presumably have to contain the explanation of its own existence, the reason for its triumph over sheer nothingness. So perhaps there was hope after all.

I WAS A little sad to leave Deutsch. Despite the gelid beginnings of our acquaintance, he had revealed a real sweetness of character and intellectual generosity. And Lulie, sitting next to me on the couch with her plate of macaroni, auditing our conversation with keen interest, her adoring eyes fixed on Deutsch, seemed a very angel. I had even grown comfortable in the towering chaos of junk that surrounded me, coming to see it as an adventure in high-entropy housekeeping.

As I made my solitary way along the highway back down to Oxford, a pinkish-orange ray of sunlight broke over the clouded horizon. The bells of the colleges were ringing in the distance again. I tried to picture myself as a denizen of Deutsch's multiverse. In innumerable parallel worlds, my quantum counterparts were also descending such a hill, were also hearing such bells, were also rejoicing in such a brilliant display of sunlight as the late-winter day waned. And, like me, they were pondering the mystery of why the multiverse exists. Their thoughts—*my* thoughts—were embodied in a physical structure that extended, like a higher-dimensional crystal, across parallel universes. Surely one of these quantum counterparts, shadowing me somewhere in Deutsch's vast fabric of reality, had made more progress than I had toward ultimate enlightenment. What thoughts could be running through *his* head? Or was the resolution to the mystery of existence somehow encoded in that crystalline structure as a whole, transcending the denizens of any particular quantum world?

Just then a passing bus startled me by hooting its horn, and my vision of this unsubstantial pageant faded, leaving not a wrack behind.

Interlude

The End of Explanation

Bertrand Russell, according to the philosophical lore, was once in the course of giving a public lecture on cosmology when he was interrupted by an old lady in the audience. "Everything you've been telling us is rubbish," the lady vociferously objected. "The world is actually flat, and it's supported by a giant elephant that is standing on the back of a turtle." Russell, humoring her, asked what might support the turtle. The old lady replied, "It's turtles all the way down!"

When it comes to understanding reality, David Deutsch turned out to be something of a "turtles all the way down" man. Our explanatory quest will be unending, he maintained. There is no bedrock principle that explains absolutely everything (including the principle itself). There is no self-supporting "superturtle" holding up the tower of turtles above.

But suppose Deutsch is wrong. Suppose there *is* an ultimate explanation for everything. What could such a principle look like? How would we know when we had reached it?

It was Aristotle, in his logical work, *Posterior Analytics*, who first addressed this matter. There are three ways an explanatory chain might go, Aristotle observed.

First, it might go in a circle: *A* is true because *B*, and *B* is true because *A*. (The circle might be widened by lots of intermediate explanatory truths: *A* because *B*, *B* because *C*, . . . *Y* because *Z*, *Z* because *A*.) But a circular explanation is no good. Saying "A because B because A" is a roundabout way of saying "A because A." And no truth explains itself.

Second, the explanatory chain might go on forever: A1 is true because A2, A2 is true because A3, A3 is true because A4, and so on, to infinity. But that's no good either. Such an endless regress, Aristotle observed, supplies no ultimate explanatory foundation for knowledge.

That leaves the third kind of explanatory chain, one that terminates in a finite number of steps: A1 because A2, A2 because A3, and so on, down to some final truth X. And what sort of truth could X be?

There would seem to be two possibilities. First, X might be a brute fact, lacking any explanation of its own. But if X itself has no explanatory support, Aristotle remarked, it can hardly provide support for other truths. The second possibility is that X is a logically necessary truth, one that could not have been otherwise. And, for Aristotle, this was the only satisfactory way for an explanatory chain to end—the only alternative to circularity, infinite regress, and unjustified explanatory danglers.

But—with due respect to Aristotle—how could a logically necessary truth really explain anything? In particular, how could it explain anything that is *logically contingent*—like the fact that there is a world? If the existence of a world could be deduced from a logically necessary truth, then it too would be logically necessary. But it isn't. Although there is a world, there might not have been. Nothingness cannot be dismissed as a logical possibility. Even the most promising attempt to derive being from pure logic—the ontological argument for the existence of God—in the end comes to nothing.

So, in our quest for total understanding, we cannot complete our explanatory chain with a logically necessary truth. We are therefore driven back to a choice among three evils: circularity, infinite regress, and brute fact. Of this trio, brute fact would appear to be the least objectionable. But is there any way the brute-fact dangler at the end of an explanatory chain can be made to seem less arbitrary? Can it be rendered less brutal?

The Harvard philosopher Robert Nozick had an interesting proposal along these lines. The only way an explanation could leave nothing at all unexplained, Nozick began by observing, is if the final truth in the series were somehow self-explanatory. But how could a truth explain itself? "X because X" is an evasion of explanation rather than the real thing. No child is satisfied if you answer the question "Why is the sky blue?" by saying "Because it is." We are back to the evil of circularity again. That is why philosophers from Aristotle to Richard Swinburne have staunchly maintained that nothing explains itself—that the explanatory relation is, to use the technical term, "irreflexive."

Nozick, however, saw more to the matter. He conceded that "X because X" is no good as an explanatory paradigm. But there is another way, he observed, that a truth might be deduced from itself. Let's say our deepest principle—the one that explained all the laws of nature—turned out to have this form:

Any law having characteristic C is true.

Let's call this deepest-of-all-principles P. The principle P explains why other laws hold true: because they have characteristic C. But what explains why P is true? Well, suppose that P turned out to have characteristic C. Then the truth of P would logically follow from P itself! In that case, principle P would be *self-subsuming*, to use Nozick's term.

"Self-subsumption is a way a principle turns back on itself, yields itself, applies to itself, refers to itself," Nozick wrote. He admitted that explanatory self-subsumption is "quite weird—a feat of legerdemain." However, compared to the alternatives—circularity, infinite regress, and brute-fact danglers—it doesn't look so bad.

Of course, showing that a principle is self-subsuming is no proof that the principle in question is valid. Consider the sentence "Every sentence of exactly eight words is true." Call this sentence S. Since S has exactly eight words, the truth of S is derivable from S itself, making it self-subsuming. But S is clearly false. (I leave this as an exercise for the reader.) Another statement that is self-subsuming yet false is "All generalizations are true."

When a self-subsuming principle *is* true, however, it does in a sense *explain* why it is true. (What is explanation, after all, but subsumption under a law?) "The ultimate principle which is true will, I have suggested, explain itself by subsuming itself," Nozick wrote. "Being a deep fact, deep enough to subsume itself and to yield itself, the principle will not be left dangling without any explanation." So as the terminus of an explanatory chain, a self-subsuming principle is certainly preferable to a brute fact.

Still, self-subsumption does not by itself eliminate all explanatory loose ends. Consider again the self-subsuming sentence S: "All sentences of exactly eight words are true." Although S is false, it is possible to imagine a world that makes it true. Even in this world, however, we would not be satisfied with S as an ultimate explanation. For one thing, it looks arbitrary. Why should S be true and not some rival self-subsuming sentence—say, "All sentences that contain exactly nine words are true"? For another thing, S does not have the look of ultimacy. If it were true, we would seek some deeper explanation of why it was the case—of why the world and language were arranged just that way.

Even though self-subsumption is not a guarantee of ultimacy, it may at least be a *mark* of ultimacy. Suppose, Nozick said, we were to find "a self-subsuming statement that is deep enough to yield everything else in an area or realm, while repeated efforts fail to find a further truth that yields it." Then, he contended, it would be "a reasonable conjecture, tentatively held and overturnable, that an ultimate truth has been reached." In other words, we may have found our superturtle.

Could some self-subsuming principle of the kind envisioned by Nozick furnish the answer to the question *Why is there something rather than nothing?* David Deutsch thought there could be no such answer, no end to explanation. Richard Swinburne thought the best we could do was to find the right explanatory "stopping point," a hypothesis of maximum simplicity and power, which for him was the existence of God. Yet Swinburne conceded that God's own existence had no explanation, "for surely never does anything explain itself." Nozick, by contrast, saw a way that a principle *could* explain itself without being blatantly circular. His ideal of self-subsumption would thus seem to mark an explanatory advance over Swinburne's ideal of simplicity.

But what sort of self-subsuming principle could explain why there is Something rather than Nothing?

Nozick thought he might have the answer. He proposed what he called "the principle of fecundity." This is the most liberal of all ontological principles. It states that *all possible worlds are real*. The principle of fecundity was not invented by Nozick. In essence, the idea—which is also known as the "principle of plenitude"—goes back to Plato. Versions of it have been entertained throughout the history of thought. What is novel with Nozick is the claim that the principle of fecundity, being self-subsuming, furnishes its own justification. "If it is a very deep fact that all possibilities obtain," he wrote, "then that fact, being a possibility, obtains in virtue of the deep fact that all possibilities do."

A reality governed by the principle of fecundity would be the richest and most expansive reality conceivable. But it would have a rather odd structure. All possible worlds would exist, but they would exist as "parallel universes," in logical isolation from one another. Some of these worlds would be very large and complicated. The largest of them, which we might call the *maximal* world, would itself contain every possibility, mirroring the richness of the entire ensemble of possible worlds that made up reality as a whole. At the other end of the range of possibilities would be the *minimal* or *null* world, representing the possibility that nothing at all existed. In between would come all possibilities of intermediate size and complexity: worlds containing a single electron and positron orbiting each other, worlds looking much like our own universe, worlds containing the Greek gods, worlds made of cream cheese, and so on.

The principle of fecundity, if true, would mean that reality is infinitely more encompassing than we had imagined. It would make our little universe look provincial in the extreme. And such a reality would have the virtue of eliminating the mystery of existence—so, at any rate, contended Nozick. The minimal world, one of the separate possibilities realized according to the principle of fecundity, is just our old friend nothingness. So why is there something rather than nothing? "There isn't," Nozick replied. "There's both."

But wait—the logic seems to have gone askew here. There can't be *both* something *and* nothing. If you have a reality that consists of bits of something and you add a bit of nothing to it, you still have something. And the absurdity does not stop there. The principle of fecundity says that all possibilities are realized. Now, one possibility is

R: Everything is red.

Another possibility is:

not-R: There is at least one thing that isn't red.

So the principle of fecundity implies *R* and not-*R*—a contradiction. And anything that implies a contradiction must be false.

Nozick had a response to this objection. Although the two possibilities *R* and *not-R* are both realized, he said, "they exist in independent noninteracting realms." We might think of them as two different planets, "Planet Red" and "Planet not-Red." That's one way out of the contradiction. But it's not a good way. For even if *R* and *not-R* prevail on separate planets, there can be no planet where both possibilities are realized together. In other words, there can be no "Planet Fecundity" among the possible planets. Even if all possible planets are realized, there is no planet where all possibilities are realized. So fecundity is not self-subsuming after all. It's a cruel dilemma for Nozick: either his ultimate explanatory principle leads to contradiction, or it fails to be self-subsuming.

A self-subsuming ultimate principle is like a barber who shaves all the men in the village and himself too. There's nothing logically wrong with that. It's the principle of fecundity that's the problem. It countenances *too many* possibilities—including the paradoxical one of a barber who shaves all and only those men who *don't* shave themselves. Given this fatal logical defect, the principle of fecundity is clearly not fit to serve as the ultimate explanation.

Is the search for a self-subsuming principle of reality then hopeless? Unhappily, Nozick himself had nothing else to offer. (He died in 2002, of stomach cancer, at the age of sixty-three.) Perhaps his ontological speculations, wild as they seemed to many of his fellow philosophers, were not quite wild enough. If philosophy, like theology before it, had so far failed to come up with the goods, maybe it was time for me to look elsewhere, in the still-wilder reaches of contemporary physics. I might not find the sought-for explanatory "superturtle" there. But I had heard theoretical physicists talking about the universe as a "free lunch," and that sounded almost as good.

8

THE ULTIMATE FREE LUNCH?

Science cannot answer the deepest questions. As soon as you ask
why there is something instead of nothing, you have gone beyond
science.

—ALLAN SANDAGE, *the father of modern astronomy*

Science is impotent to address the mystery of existence—so, at least, it
is often claimed. The point was put forcefully by the secular humanist (and evolutionary biologist) Julian Huxley. "The clear light of
science, we are often told, has abolished mystery, leaving only logic and
reason," Huxley wrote. "This is quite untrue. Science has removed the
obscuring veil of mystery from many phenomena, much to the benefit of
the human race: but it confronts us with a basic and universal mystery—the
mystery of existence.... Why does the world exist? Why is the world-stuff
what it is? Why does it have mental or subjective aspects as well as material
or objective ones? We do not know.... But we must learn to accept it, and
to accept its and our existence as the one basic mystery."

The question *Why is there something rather than nothing?* is supposed to
be "too big" for science to explain. Scientists can account for the organization of the physical universe. They can trace how the individual things
and forces within it causally interact. They can shed light on how the universe as a whole has, in the course of its history, evolved from one state
into another. But when it comes to the ultimate origin of reality, they have
nothing to say. That is an enigma best left to metaphysics, or to theology,
or to poetic wonderment, or to silence.

As long as the universe was thought to be eternal, its existence did not greatly vex scientists anyway. Einstein in this theorizing simply assumed the eternity of the universe, and he fudged his relativity equations accordingly. With the discovery of the Big Bang, however, everything changed. We are evidently living in the dilute, expanding, cooled-down remnants of a great cosmic explosion that occurred some 14 billion years ago. What could have caused this primal explosion? And what, if anything, preceded it? These certainly sound like scientific questions. But any attempt by science to answer them faces a seemingly insuperable obstacle, known as the *singularity*.

Suppose we take the laws of general relativity, which govern cosmic evolution on the largest scale, and extrapolate them backward in time toward the beginning of the universe. As we watch the evolution of our expanding and cooling cosmos in reverse, we would see its contents contracting and growing hotter. At $t = 0$—the moment of the Big Bang—the temperature, density, and curvature of the universe all go to infinity. Here the equations of relativity break down, become meaningless. We have reached a singularity, a boundary or edge to spacetime itself, a point at which all causal lines converge. If there *is* a cause for this event, it must transcend spacetime and hence escape the reach of science.

The conceptual breakdown of science at the Big Bang was disturbing to cosmologists, so disturbing that they searched for scenarios in which the initial singularity was somehow avoided. But in 1970, the physicists Stephen Hawking and Roger Penrose showed that these efforts were futile. Hawking and Penrose began by assuming, quite reasonably, that gravity is always attractive, and that the density of matter in the universe is roughly what it has been measured to be. Given this pair of assumptions, they proceeded to prove, with mathematical certainty, that there *must* have been a singularity at the beginning of the universe.

Did this mean that the ultimate origin of the universe is forever shrouded in unknowability? Not necessarily. It merely means that the Big Bang cannot be completely understood by "classical" cosmology—that is, the kind of cosmology that is based on Einstein's general relativity alone. Other theoretical resources would be needed.

As a clue to what kind of resources, consider that, a fraction of a second after its birth, the entire observable universe was no bigger than an atom. At that size scale, classical physics no longer applies. It is quantum theory that governs the realm of the very tiny. So cosmologists—Stephen Hawking prominently among them—began to ask, What if quantum theory, previously used to describe subatomic phenomena, were applied to the universe as a whole? Thus was born the field of *quantum cosmology*, which has been described (by the physicist John Gribbin) as "the most profound development in science since Isaac Newton."

Quantum cosmology seemed to offer a way around the singularity problem. Classical cosmologists had supposed that the singularity lurking behind the Big Bang was a pointlike thing, with zero volume. But quantum theory forbids such a sharply defined state of affairs. It decrees that nature, at the most fundamental level, is irredeemably fuzzy. It rules out the possibility of a precise temporal origin to the universe, a time $t = 0$.

But what is more interesting than what it *forbids* is what quantum theory *permits*. It permits particles to pop into existence spontaneously, if briefly, out of a vacuum. This scenario of creation *ex nihilo* led quantum cosmologists to entertain an arresting possibility: that the universe itself, through the laws of quantum mechanics, bounded into existence out of nothing. The reason there is Something rather than Nothing is, as they fancifully put it, that *nothingness is unstable.*

The physicist's statement "nothingness is unstable" is sometimes mocked by philosophers as an abuse of language. "Nothingness" does not name an object, they say; therefore, it is meaningless to ascribe a property, like instability, to it. But there is another way of thinking of nothingness: not as a thing, but as a description of a state of affairs. For a physicist, "nothingness" describes a state of affairs where there are no particles and where all the mathematical fields have the value zero.

Now we can ask, Is such a state of nothingness possible? That is, is it logically consistent with physical principles? One of the deepest of these principles, lying at the very basis of our quantum understanding of nature, is Heisenberg's uncertainty principle. This principle says that certain pairs of properties—called "canonically conjugate variables"—are linked in such a way that they cannot both be measured precisely. One such pair is position and momentum: the more precisely you locate the position of a particle, the less you know about its momentum, and vice versa. Another pair of conjugate properties is time and energy: the more precisely you know the time span in which something occurred, the less you know about the energy involved, and vice versa.

Quantum uncertainty also forbids the precise determination of the *value* of a field and the *rate* at which that field value changes. (That's like saying you can't know the exact price of a stock and how quickly that price is changing.) And, when you think about it, this pretty much rules out nothingness. Nothingness is, by definition, a state in which all field values are timelessly equal to zero. But Heisenberg's principle tells us that if the value of a field is precisely known, its rate of change is completely random. In other words, that rate of change can't be precisely zero. So a mathematical description of changeless emptiness is incompatible with quantum mechanics. To put the point more pithily, nothingness is unstable.

Could this have something to do with cosmogenesis? The thought that it might seems first to have occurred back in 1969 to a New York City physicist named Ed Tryon. Doing a bit of wool-gathering during a talk by a visiting celebrity physicist at Columbia University, Tryon suddenly blurted out, "Maybe the universe is a quantum fluctuation!" The remark was reportedly greeted by the several Nobel laureates present with derisive laughter.

But Tryon was on to something. It may seem implausible that a universe containing so much sheer stuff—there are a hundred billion galaxies just in the little region of it we can observe, each with a hundred billion stars—could have arisen from *nothing*. As we know from Einstein, all of this mass is frozen energy. But against the vast amount of *positive* energy locked up in the stars and galaxies must be set the *negative* energy of the gravitational attraction among them. In fact, in a "closed" universe—one that will eventually collapse back on itself—these positive and negative energies precisely cancel each other. In other words, the net energy of such a universe is zero.

The possibility that the entire universe could be made out of no energy at all is an astonishing one. It certainly astonished Einstein: when the idea was explained to him by a fellow physicist, George Gamow, while the two were walking in Princeton, a stunned Einstein "stopped in his tracks," Gamow recalled, "and, since we were crossing a street, several cars had to stop to avoid running us down."

From the quantum point of view, a zero-energy universe presents an interesting possibility, which Tryon seized upon. Suppose the total energy of the universe is indeed exactly zero. Then, owing to the trade-off in uncertainty between energy and time (as decreed by the Heisenberg principle), the indeterminacy in its time span becomes infinite. In other words, such a universe, once it popped into existence out of the void, could run away with itself and last forever. It would be like a loan of pure being that need never be repaid. As for what "caused" such a universe to pop into existence, that is simply a matter of quantum chance. "In answer to the question of why it happened," Tryon later commented, "I offer the modest proposal that our universe is simply one of those things which happen from time to time."

Is this *creatio ex nihilo*? Not quite. It is true that Tryon's genesis scenario has a zero cost in terms of energy and matter; in that sense, it does seem to get "something from nothing." But the state out of which Tryon's cosmos spontaneously materializes, called the "quantum vacuum," is very far from the philosopher's conception of nothingness. For one thing, it is a kind of empty *space*, and space is not nothing. Nor is the space of the quantum vacuum really empty. It has a complicated mathematical structure; it bends and flexes like rubber; it is saturated with energy fields and seethes with virtual-particle activity. The quantum vacuum is a physical object; indeed, it is a little proto-cosmos unto itself. Why should such a thing as a quantum vacuum ever have existed? As the physicist Alan Guth has observed, "A proposal that the universe was created from empty space seems no more fundamental than a proposal that the universe was spawned by a piece of rubber. It might be true, but one would still want to ask where the piece of rubber came from."

The man who seems to have come the closest to solving the "rubber problem" is Alex Vilenkin. Vilenkin was born in Ukraine, in the former

Soviet Union, where, after obtaining an undergraduate degree in physics, he held a job as a night watchman in a zoo. In 1976 he immigrated to the United States, and in little more than a year he managed to earn a Ph.D. in physics. Vilenkin now teaches at Tufts University near Boston, where he is also director of the Tufts Institute of Cosmology. He is known for wearing dark glasses during seminars, Anna Wintour–like, supposedly because of the sensitivity of his eyes to light.

When Vilenkin talks about the universe arising from "nothing," he means it quite literally, as I learned from chatting with him a few years ago. "Nothing is nothing!" he insisted to me, with some vehemence. "Not just no matter. It's no space. No time. *Nothing.*"

But how could a physicist even define a state of sheer nothingness? Here is where Vilenkin showed ingenuity. Imagine spacetime as the surface of a sphere. (Such a spacetime is called "closed," since it curves back on itself; it is finite, even though it has no boundaries.) Now suppose that this sphere is shrinking, like a balloon that is losing its air. The radius grows smaller and smaller. Eventually—try to imagine this—the radius goes all the way to zero. The surface of the sphere disappears completely, and with it spacetime itself. We have arrived at nothingness. We have also arrived at a precise *definition* of nothingness: a closed spacetime of zero radius. This is the most complete and utter nothingness that scientific concepts can capture. It is mathematically devoid not only of stuff but also of location and duration.

With this characterization in hand, Vilenkin was able to do an interesting calculation. Using the principles of quantum theory, he showed that, out of such an initial state of nothingness, a tiny bit of energy-filled vacuum could spontaneously "tunnel" into existence. How tiny would this bit of vacuum be? Perhaps as little as one hundred-trillionth of a centimeter. But that, it turns out, is good enough for cosmogonic purposes. Driven by the negative pressure of "inflation," this bit of energetic vacuum would undergo a runaway expansion. In a couple of microseconds it would attain cosmic proportions, issuing in a cascading fireball of light and matter—the Big Bang!

So the transition from Nothingness to Being, as imagined in the

Vilenkin scenario, is a two-stage affair. In the first stage, a tiny bit of vacuum appears out of nothing at all. In the second stage, this bit of vacuum
blows up into a matter-filled precursor of the universe we see around us
today. The whole scheme would appear to be scientifically irreproachable.
The principles of quantum mechanics, which govern the first stage, have
so far proved to be the most reliable principles in all of science. And the
theory of inflation, which describes the second stage, not only has been a
conceptual success since it was introduced in the early 1980s, but also has
been triumphantly confirmed by empirical observations—notably, by the
patterns of background radiation left over from the Big Bang that have
been observed by the COBE satellite.

So Vilenkin's calculations appeared to be sound. Yet, in chatting with
him, I had to confess that my imagination bridled at his scenario of creation from nothing. Surely, the bubble of false vacuum out of which the
cosmos was born had to come from *somewhere*. So, rather impishly, he
told me to picture the bubble forming in a glass of champagne—and then
to subtract the champagne.

Even with this image in mind—a not-altogether persuasive one—I
remained perplexed. A champagne bubble forms in the course of time.
But Vilenkin's bubble that appears out of nothingness is a bubble of *spacetime*. Since time itself (along with space) is created in the transition from
Nothing to Something, this transition can't very well take place *in* time.
It seems to unfold logically rather than temporally. If Vilenkin is right,
nothingness never had a chance: the laws of physics eternally ordained
that, with some appreciable probability, there would be a universe. But
what gives ontological clout to these laws? If they are logically prior to the
world, where exactly are they written down?

"If you like," Vilenkin told me, "you can say they're in the mind of God."

Perhaps, I thought after talking with Vilenkin, this is the best that science could do. It could show that the laws explaining how things happen
within the world also explain why there should *be a world at all*—and hence,
why there is Something rather than Nothing. The laws of classical physics,
including those of Einstein's general relativity, were not up to this challenge. They could describe the evolution of the universe, but they could

not account for its coming into being; indeed, they broke down at its point of origin. Quantum cosmology was an improvement. It could treat the origin of the world as just another quantum event, one mercifully free from the need for a First Cause. It could show that, ontologically speaking, the universe might indeed be a "free lunch."

But quantum cosmology cannot, scientifically speaking, be the last word. The problem is that, as yet, no one has been able to explain how *gravity* fits into the quantum framework. Gravity is, after all, the force of nature that determines the overall architecture of the universe. On such a large scale, Einstein's general relativity suffices to explain the workings of gravity. But when the entire mass of the universe is packed into a volume the size of an atom—as was the case just after the Big Bang—quantum uncertainty causes the smooth geometry of general relativity to break up, and there is no telling how gravity will behave. To understand the birth of the cosmos, we thus need a quantum theory of gravity, one that "unifies" general relativity and quantum mechanics. Stephen Hawking himself has conceded as much. "A quantum theory of gravity is essential if we are to describe the early universe," Hawking declared in 1980, in his inaugural lecture as the Lucasian Professor of Mathematics at Cambridge University. "Such a theory is also required," he said, "if we are to answer the question: Does time really have a beginning?"

Today, more than three decades later, physicists are still searching for a theory of the type that Hawking had in mind, one that would tie up all the forces of nature—including gravity—into a single neat mathematical package. It is not yet clear what form this final theory will take. At the moment, the physics community is pinning its hopes on "string theory," which seeks to interpret all of physical reality as consisting of tiny strings of energy vibrating in higher-dimensional space. Dissenters from the string-theoretic consensus are trying other approaches. And a few physicists think the whole idea of unification is a pipe dream.

What might a final theory—or a "Theory of Everything," as it's sometimes called—tell us about the origin of the universe? Such a theory is likely to go deeper than the quantum cosmology of Hawking, Vilenkin, and company. (String theory, for example, offers glimpses of a pre–Big Bang reality where the very notions of space and time have no purchase.) But could it provide a convincing rationale for the world's existence? And could it provide a convincing rationale for *itself*? If it is truly a Theory of Everything, it ought to be able to explain why it is itself true. Could the Theory of Everything turn out to be self-subsuming?

The thinker in the best position to answer such questions, I knew, was Steven Weinberg. No physicist has been more central to the quest for a final theory. In 1979, Weinberg was awarded the Nobel Prize in Physics for his role, a decade earlier, in unifying two of the four fundamental forces of nature: the electromagnetic force and the "weak" force (responsible for radioactive decay). Both these forces, his work helped to show, were merely low-energy aspects of a more basic "electroweak" force. For this and related achievements, Weinberg has a good claim to be the father of the "standard model" of particle physics, which represents the most complete understanding we have of the physical world at the micro-level.

Weinberg is also an exceptionally eloquent expositor of science. In 1977 he published *The First Three Minutes*, a cinematically gripping account of the primeval universe in the explosive moments after the Big Bang. (It was on the last page of this book that he made his to-be-notorious declaration, "The more the universe seems comprehensible, the more it also seems pointless.") In 1993 he published *Dreams of a Final Theory*, which explained, with real philosophical profundity, what was at stake in the quest to unify the laws of nature. Weinberg described how physicists, guided by their sense of mathematical beauty, sought deeper and deeper principles that would merge the standard model and Einstein's general relativity into an all-encompassing final theory. This would be a point where all the arrows of explanation converge—where every *why* is absorbed in an ultimate *because*. Weinberg explained why he thought that contemporary physics might be on the verge of discovering just such a theory. He even confessed to some sadness at the prospect, writing that "with the discovery of a final theory we may regret that nature has become more ordinary, less full of wonder and mystery."

Just how much cosmic mystery did Weinberg think the final theory would leave as a residue? He was quite explicit in denying that it could explain literally *everything*. For example, Weinberg did not think that science could ever explain the existence of moral truths, owing to the logical gap between the scientific *is* and the ethical *ought*. But could science explain the existence of the world? Could it account for the triumph of Something over Nothing?

I was eager to put such questions to Weinberg. In fact, I was eager to meet him, period. There was no other living physicist I regarded with such awe. And there was no other physicist (with the exception of Freeman Dyson) who had such a gift for putting his ideas in lapidary form. Besides, Weinberg seemed to be an extraordinary-looking fellow, judging from descriptions of him I had seen in the press. "With his crab-apple cheeks, vaguely Asian eyes, and silver hair still tinged with red, Steven Weinberg resembles a large, dignified elf," one journalist wrote of him after an encounter. "He would make an excellent Oberon, king of the fairies in *A Midsummer Night's Dream*."

So, feeling rather like Nick Bottom myself, I got in touch with Weinberg. He teaches at the University of Texas in Austin, having come there in 1982 after previously holding the Higgins Chair of Physics at Harvard. I proposed making a pilgrimage to Austin to talk to him about the mystery of existence. He responded graciously to this threatened imposition on his time. "If you are coming all the way down here from New York, I'll even buy you lunch," he wrote in an e-mail. The universe, I thought, was not the only free lunch.

The prospect of visiting Austin for the first time was an added allurement to me. From what I had heard of the place, I pictured it as a wondrous bastion of avant-garde culture and bohemian living in an otherwise medievally backward state. It even seemed to be theologically progressive. When I had asked Weinberg, who has inveighed against religion ("With or without religion, you would have good people doing good things and evil people doing evil things. But for good people to do evil things, that takes religion."), how he could be happy in a Baptist hotbed like Texas, he assured me that, far from being uniformly fundamentalist, some of the Baptist congregations there were so liberal as to be indistinguishable from Unitarians. And I was impressed by Austin's reputation as the live-music capital of the world, even if indie rock was not exactly my dish.

So I eagerly booked a flight to Austin and made a reservation at the Intercontinental Hotel for what promised to be an intellectually stimulating and altogether delightful weekend—not realizing that my plans were about to be destroyed by a little eruption of *le néant* into my life.

Interlude

Nausea

I t was early afternoon on Saturday when my plane touched down at the Austin airport. For a late-spring day, the heat and humidity were surprising, and even in my linen suit, elegantly rumpled as always, I felt a little uncomfortable.

On the way downtown I noticed a lot of activity in the streets. It seemed that some sort of outdoor music festival was getting under way.

After checking into my hotel, I went for a stroll around the old downtown. By now the music festival was in full swing. Rockabilly garage bands blared away on every block; beery throngs pressed in and out of bars; meat sizzled on grills in the middle of the closed-off streets. The noise was intense. So were the smells.

Making my way through the cacophonous crush under the hot sun, I pretended that I was Roquentin, the existential hero of Sartre's novel *Nausea*. I tried to summon up the disgust he would feel at the surfeit of Being that overflowed the streets of Austin—at its sticky thickness, its grossness, its absurd contingency. Whence did it all spring? How did the ignoble mess around me triumph over pristine Nothingness? Roquentin, overwhelmed by the gelatinous slither of existence that environed him in his lonely wanderings through Bouville, was moved to shout, "Filth! What rotten filth!" I might have done the same, but my epiphany was too feeble to justify such an anguished outburst. Besides, everyone around me seemed to be having an awfully good time.

By evening, the streets of Austin had quieted down a little. I asked the concierge at my hotel for advice on where to dine. He recommended a restaurant called the Shoreline Grill, which was located next to Lady Bird Lake, a river-like body of water, apparently named for the late wife of President Lyndon Johnson, that ran through the city.

When I got to the restaurant, I encountered a group of high-school students arrayed in formal wear. It was prom night in Austin, and they were having a fancy dinner there before going to the dance. As I was to discover a few weeks later, Steven Weinberg also happened to be dining at the Shoreline that night, in another room from the one where the maitre d' seated me. That, as events fell out, was as close as our world-lines came to intersecting.

It was only dusk when I finished my meal in the midst of the prom-goers. On leaving the restaurant, I noticed a large and relatively silent crowd of people assembled by a bridge that spanned Lady Bird Lake. They seemed to be waiting for something. I asked one of them what was going on. He pointed under the bridge. "Bats," he said in a hushed voice. "They're all going to take off together in just a few minutes. Happens every night. It's something to see."

Looking more closely at the dark underside of the bridge, I saw that it was a continuous carpet of hanging bats—more than a million of them, I was told. They were "Mexican free-tailed bats." On nice evenings, like this one, tourists and locals would line the lake shore to await the dramatic moment when the bats, ravenous for their nightly meal of insects, took wing in a single giant swarm, blackening the sky.

Having nothing better to do, I sat down on the grassy bank of the lake and waited along with them. The minutes passed. The bats did not stir. A boat chugged by. More minutes passed. Still the bats did not stir. It grew dark. Disappointed, the crowd began to drift away. I got up off the grass and walked back to my hotel, thinking that this unfulfillment was not a good omen for my meeting the next day with Weinberg.

When I entered my room, I noticed the light on the phone was blinking. Someone had left a message. It turned out to be the couple who were taking care of my dog, a little long-haired dachshund named Renzo, during my absence from New York. I called them immediately. Renzo, they gravely told me, had had some kind of seizure earlier that day. While frolicking in the chicken run of their weekend farm in rural Pennsylvania, he had suddenly collapsed with a howl. They had wrapped his semi-comatose body in a cold wet towel and driven him to the emergency room of a nearby animal hospital.

I imagined Renzo alone in a dark and unfamiliar kennel, possibly dying, and wondering, in his flickering consciousness, where I was. There was no choice. After an hour or so of haggling with various airlines, I had arranged to fly back to New York first thing in the morning. I sent a regret-filled e-mail to Weinberg, telling him that a "family emergency" had intervened to make our lunch the next day impossible. Then I dropped into bed and had a fitful night of sleep as the noisy air-conditioning in my room cycled on and off.

When I called the animal hospital the next morning, they told me that Renzo seemed better. He had eaten a little food and had even tried to bite one of the vets. Cheered by this news, I managed to endure the tedious sequence of connecting flights home. But when I was reunited with my dog at the end of that long day, my optimism vanished. Something was terribly wrong with him.

Subsequent X-rays confirmed my worst fears. Renzo's lungs and liver, the vet told me, showed signs of cancer. The cancer had probably metastasized to his brain, causing the seizure. He seemed to have lost his sight and sense of smell, which suggested that the parts of his cortex responsible for visual and olfactory processing had been destroyed.

Renzo's once-rich canine sensory world had disappeared into nothingness. All he could do was blindly stumble around in circles, whimpering in distress. Only when I held him in my arms did he seem to get some relief.

So I spent the next ten days holding him. Occasionally he would lick my hand or even wag his tail a little. But his condition was clearly getting worse. He stopped eating. He was unable to sleep, crying through the night in pain. When even the strongest painkillers would not abate his agony, I knew it was time for the inevitable.

I stayed in the room with my dog during his euthanasia. The process took about a half hour. First, Renzo was given a tranquilizing shot. This caused his writhing and whimpering to stop. Stretched out on the table, at peace for the first time in days, he suddenly looked much younger than his fourteen years. He was breathing slowly, and his eyes, though sightless, were open. Then a catheter was inserted into his paw for the lethal injection.

The vet in charge of all this looked like a young Goldie Hawn. She and her assistant took turns with me stroking Renzo during the preparations. I did not want to break down sobbing in front of them.

Fortunately, I have a good trick for maintaining my outward composure in such situations. It involves a beautiful little theorem about prime numbers, originally due to Fermat. Pick a prime number—13, for example. See if it leaves a remainder of 1 when divided by 4. If it passes this test—as 13 does—then, says the theorem, that prime number can always be expressed as the sum of two squares. And sure enough, $13 = 4 + 9$, each of which is a square. My trick for controlling myself in moments of unbearable emotion is to run through the numbers in my head and apply this theorem to each one in turn. First, I check to see whether the number is a prime that leaves a remainder of 1 when divided by 4; if it is, I mentally break it down into two squares. For the smaller numbers, this is easy. It's immediately apparent, for example, that 29 is a prime number that leaves a remainder of 1 when divided by 4, and it's also easy to see that 29 is the sum of the two squares 4 and 25. When you get past 100, though, both tasks become more challenging if you don't have pencil and paper. Take the number 193. You have to poke it a bit to make sure that it is indeed the right kind of prime for the theorem to apply. And once you have done this, it may take more than a few seconds to see that the squares it breaks up into are 49 plus 144.

I had made it past 193 and was still dry-eyed at the moment the vet gave Renzo the final injection, the one that would paralyze his nervous system and shut down his little heart. It did its work quickly. Just a moment after the plunger was fully depressed, he exhaled in a burst. "That was his last breath," the vet said. Then he exhaled again, and was still. Good dog.

The vet and her assistant left me alone in the room so that I could sit for a while with Renzo's lifeless body. I opened his mouth and looked at his teeth, something he would never let me do when he was alive. I tried to close his eyes. After a few more minutes, I left the room and paid the bill, which included a "communal cremation" with other dogs that had been put down. Then, carrying only Renzo's blanket, I walked home.

The next day, I called Steven Weinberg at his home in Austin to talk about why the world exists.

9

WAITING FOR THE FINAL THEORY

"So you didn't like the Shoreline Grill? I thought the food there was reasonably good. Pricey for Austin, but not by New York standards. By the way, I've completely forgotten why we're having this chat."

It was Steven Weinberg, speaking over the phone in his deeply sonorous, ironically gruff voice.

I reminded him that I was writing about why there is Something rather than Nothing.

"That's a nice idea for a book," he said, his tone rising on the word "nice."

The compliment was gratifying. But did he feel the same way that Wittgenstein and so many others did about the question? Was he awed at the very fact of existence? Did he find it extraordinary that there should be a world at all?

"For me," Weinberg said, "it's part of a larger question, which is 'Why are things the way they are?' That's what we scientists try to find out, in terms of deep laws. We don't yet have what I call a final theory. When we do, it might shed some light on the question of why there is anything at all. The laws of nature might dictate that there *has* to be something. For example, those laws might not allow for empty space as a stable state. But that wouldn't take away the wonder. You'd still have to ask, 'Why are the laws *that* way, rather than some *other* way?' I think we're permanently doomed to that sense of mystery. And I don't think belief in God helps. I've said it before, and I'll repeat it. If by 'God' you have something definite in mind—a being that is loving, or jealous, or whatever—then you're faced with the question of why God's that way and not another way. And if you don't have anything very definite in mind when you talk about 'God' being behind the existence of the universe, then why even use the word? So I think religion doesn't help. It's part of the human tragedy: we're faced with a mystery we can't understand."

And Weinberg didn't seem to think that his fellow physicists could shed much light on the ultimate origin of the universe either. "I'm very skeptical," he said, "because we don't really understand the physics. General relativity breaks down when you go back to extreme conditions of density and temperature near the Big Bang. I'm also skeptical of anyone who quotes theorems about inevitable singularities—Hawking theorems and so on. Those theorems are valuable because they imply that at a certain point in, say, the collapse of a star our theories don't apply any more. But beyond that you can't say anything. We're just too ignorant at the moment."

This epistemic modesty was refreshing after all the wild speculation I'd been hearing over the past year. I felt I was talking to a latter-day Montaigne, or Socrates. But what did Weinberg think of the efforts of some of his more adventurous peers to explain existence itself? I mentioned Alex Vilenkin's notion that the present universe might have inflated out of a little nugget of "false vacuum" that itself "quantum-tunneled" into being from sheer nothingness. Physics or metaphysics?

"Vilenkin is a really clever guy, and these are fascinating conjectures," Weinberg said. "The problem is that we have no way, at present, of deciding whether they're *true* or not. It's not just that we don't have the observational data—we don't even have the *theory*."

When we do have the theory—the final theory of physics—that would furnish the last word, scientifically speaking, on how the universe came into existence. But would it also explain *why* the universe exists?

"We don't know," Weinberg said. "It depends what the final theory ends up looking like. Suppose it looks like Newton's theory. In Newton's theory, there's a clear separation between *laws* and *initial conditions*. For instance,

Newtonian physics offers no hint about the initial conditions of the solar system. Newton himself was aware of this—he thought the initial conditions were laid down by God."

If the final theory allows for unexplained initial conditions—sometimes called "boundary conditions"—then even if it can fully account for the evolution of the universe, it will still leave the origins of the universe cloaked in mystery. Who or what decreed *those* initial conditions? I thought of one of the "messages from the unseen" that the great Alan Turing left behind at his death: *Science is a differential equation. Religion is a boundary condition.*

"If the final theory turned out to be like that, I'd be disappointed," Weinberg continued. "Hawking and others hope that the final theory will fix all the initial conditions, that it will leave no freedom to the universe as to how it began. But we just don't know yet."

Well, I said, let's be optimistic. Let's assume the final theory *will* account for everything about the universe, including its initial conditions. That would still leave open the question of why that final theory takes the particular form it does. Why should it describe a world of quantum particles interacting through certain forces? Or a world of vibrating strings of energy? Or any world at all? Clearly, the final theory won't be dictated by logic alone. There is more than one logically consistent way that reality could have turned out to be. But maybe there is only one logically consistent final theory that describes a reality rich enough to include conscious observers like us.

"That would be really interesting," Weinberg said. "Would it be cause for wonder? I just had a correspondence with a philosopher at Cornell about the so-called anthropic principle. This philosopher thought, if I understood him correctly, that the universe *had* to be such as to allow for observers to evolve within it—in other words, that a universe without conscious observers would be logically inconsistent. So he wasn't surprised that this universe seemed improbably fine-tuned for life. In me, this apparent fine-tuning arouses wonder. The only explanation for it, other than a theological explanation, is in terms of a *multiverse*—I mean a universe consisting of many parts, each with different laws of nature and different values for its constants, like the 'cosmological constant' which governs cosmic expansion. If there *is* a multiverse consisting of many universes, most of them hostile to life but a few favorable to it, then it's not surprising that we find ourselves in one where conditions are in the fortunate range."

Still, I observed, that would leave open the question of why this huge ensemble of universes should exist.

"I'm not saying that the multiverse would resolve *all* philosophical issues. It would eliminate the sense of wonder surrounding the fact that conditions in our universe are just right for life and consciousness. But we would still be faced with the mystery of why the laws of nature are such as to produce the multiverse of which our universe is a part. And I don't see any way out of *that* mystery. Believing that a theory can bring a world into existence is a little like believing in Saint Anselm's ontological proof of the existence of God. Anselm asks, can you think of something than which nothing more perfect can be conceived? If you're stupid enough to say yes, then Anselm goes on to show you that since existence is a perfection, it follows that the being you're thinking of must exist, because if it didn't, you could conceive of something more perfect: the same being, only existent! The ontological proof has been shot down and resurrected many times. There's a modern theologian at Notre Dame named Alvin Plantinga who claims to have a version of it which is watertight. I think it's all nonsense myself. It seems obvious to me that you can't go from thinking about something to concluding that it exists. And it also seems obvious to me that the laws of nature cannot *require* that they describe something real. No theory can tell you that the things it talks about exist."

Maybe then, I said, quantum theory holds out the best hope for an explanation of existence. It not only explains events *in* the world. It also— unlike the classical physics that it overturned—offers to explain the event of that world's coming into existence in the first place. By quantum uncertainty, it says, a seed for the cosmos is bound to pop out of the void. So the same theory that works inside the world might also buttress the world's existence from the outside.

"Yes, that might be a plus in its favor," Weinberg said. "But there's something that I'm not very happy about here. Quantum mechanics is really an empty stage. It doesn't tell you anything by itself. That's why I think Karl Popper was wrong to say that a scientific theory must be open to falsification. You can't falsify quantum mechanics, since it doesn't make predictions. It's a very general framework, one within which you can formulate theories that *do* make predictions. Newtonian physics isn't formulated in quantum mechanics, but all our modern theories are. And quantum mechanics by itself does not say anything about the universe spontaneously coming into existence. For that sort of thing, you need quantum mechanics with other theories married to it."

So where does that leave us?

"In a fairly unsatisfactory place, I'd say. In the long run, one would like to have a truly unified theory—not just quantum mechanics plus stuff, but a theory that combines everything into an indissoluble union. And nothing we've seen so far is at all like that. I mean, you can have a quantum theory of gravity, or quantum electrodynamics, or the standard model, but that's just adding players to the quantum stage. We still seem to be far from the final theory."

When I brought up string theory, a melancholy strain became detectable in Weinberg's voice.

"I was hoping that with string theory things would fall into place much more rapidly than they have," he said. "But it's been rather disappointing. I'm not one of those people who bad-mouth string theory. I still think it's the best effort we've made to step beyond what we already know, but it hasn't worked out the way we were expecting it would. There are an enormous number of different solutions to the equations of string theory, something like 10 to the five-hundredth power. If each of these solutions is somehow realized in nature, then string theory would provide a natural multiverse, and a pretty big one—big enough for the anthropic principle to work very nicely."

Weinberg was referring to what string theorists call "the Landscape": an inconceivably vast ensemble of "pocket universes," each embodying a different possible solution to the equations of string theory. These pocket universes would vary in the most basic ways: in the number of spatial dimensions they had, in the kind of particles of which their matter consisted, in the strengths of their forces, and so on. Most of them would be bio-unfriendly "dead universes," devoid of life or consciousness. But a few among this enormous multiplicity would be bound to have just the right features for intelligent observers to emerge—observers who would then be amazed to find themselves in a world that seemed miraculously fine-tuned for their comfort. Some physicists find this string-theoretic vision of the Landscape thrilling. Others contemptuously view it as a *reductio ad absurdum*.

"By the way," Weinberg added, "there's another approach to the multiverse, one that's purely philosophical. Robert Nozick, a philosopher at Harvard—he's dead now—came up with it. Nozick thought it was a philosophical principle that everything you can imagine existing actually does exist."

Right, I said—the "principle of fecundity."

"Exactly. So, on Nozick's picture, there are all of these different possible worlds, all causally disconnected from one another, each subject to entirely different laws. There's a world in which Newton's mechanics apply, and another where there are just two particles orbiting around each other forever, and still another that is totally empty. You can justify the principle of fecundity, as Nozick did, by pointing out that it has a certain pleasing self-consistency. The principle says that all possibilities are realized, but the principle itself is just one of those possibilities, so by its own lights it must be realized."

I objected that the principle of fecundity, far from being self-consistent, might be so ontologically prodigal that it actually leads to contradiction. It's like the set of all sets—which, being a set, has to contain itself. But if some sets contain themselves, one can also consider the set of all sets that *don't* contain themselves. Call this set *R*. Now ask, Does *R* contain itself? If it does, then by definition it doesn't; and if it doesn't, then by definition it does. Contradiction! (Weinberg, of course, immediately recognized this as Russell's paradox.) The fecundity principle, I claimed, suffered from a similarly fatal logical defect. If all possibilities are realized, and some possibilities include themselves, whereas others don't, then the possibility that all self-excluding possibilities are realized must itself be realized.

And that possibility is as self-contradictory as the set of all sets that do not contain themselves.

This led to an extended argument between Weinberg and me over just what it means for one possibility to exclude another. The argument ended somewhat inconclusively when we both agreed that it amounted to no more than "metaphysical fun." After a little light chat about life in New York—Weinberg was born here to immigrant parents in 1933 and attended the Bronx High School of Science, but confessed that he hadn't been back to the city "for years"—my conversation with the father of the standard model of physics was over.

HAD IT DEEPENED my insight into the mystery of existence? Well, I was surprised that Weinberg, so bracingly skeptical and scientifically tough-minded, had declared himself open to a metaphysically extravagant notion like the principle of fecundity. I went back to his *Dreams of a Final Theory* to see what he might have to say there about the matter. The principle of fecundity, he had written, "supposes that there are entirely different universes, subject to entirely different laws. But, if these other universes are totally inaccessible and unknowable, then the statement that they exist would seem to have no consequences, except to avoid the question of why they do not exist. The problem seems to be that we are trying to be logical about a question that is not really susceptible to logical argument: the question of what should or should not engage our sense of wonder."

The best that physicists can do to satisfy this sense of wonder, Weinberg seemed to believe, is to discover their holy grail, the final theory. "This may happen in a century or two," he has written, "and if it does then I think that physicists will be at the extreme limits of their powers of explanation."

The final theory envisaged by Weinberg promises to go far beyond current physics in clarifying the origins of the universe. It might, for instance, show how space and time emerged from still more fundamental entities that we as yet have no conception of. But it is hard to see how even a final theory could explain why there is a universe instead of nothing at all. Are the laws of physics somehow to inform the Abyss that it is pregnant with

Being? If so, where do the laws themselves live? Do they hover over the world like the mind of God, commanding to exist? Or do they inhere within the world, amounting to a mere summary of what goes on inside it?

Cosmologists like Stephen Hawking and Alex Vilenkin sometimes entertain the first possibility, only to be perplexed by it. Here, for example, is Vilenkin on the "quantum tunneling" by which, he submits, the universe might have been born from nothing at all: "The tunneling process is governed by the same fundamental laws that describe the subsequent evolution of the universe. It follows that the laws should be 'there' even prior to the universe itself. Does this mean that the laws are not mere descriptions of reality and can have an independent existence of their own? In the absence of space, time, and matter, what tablets could they be written upon? The laws are expressed in the form of mathematical equations. If the medium of mathematics is the mind, does this mean that mind should predate the universe?" As to *whose* mind this might be, Vilenkin passes over that question in silence.

Hawking, too, has admitted to bafflement over the ontological status, and seeming potency, of the laws of physics: "What is it that breathes fire into the equations and makes a universe for them to govern? Is the ultimate unified theory so compelling that it brings about its own existence?"

If the ultimate laws of physics, like Plato's eternal and transcendent Forms, did have a reality of their own, that would only raise a new mystery— two mysteries, in fact. The first is the one that bothered Hawking. What gives these laws their ontic clout, their "fire"? How do they reach out and make a world? How do they force events to *obey* them? Even Plato needed a divine craftsman, a "demiurge," to do the actual work of fashioning the world according to the blueprint that the Forms provided.

The second mystery that arises if the laws of physics have their own transcendent reality is even more basic: Why should those laws exist? Why not some other set of laws or, even simpler, no laws at all? If the laws of physics are Something, then they cannot explain why there is Something rather than Nothing, since they are a part of the Something to be explained.

So consider the other possibility—that the laws of physics have *no* ontological status of their own. On this view, these laws do not hover over the world or exist prior to it in any way. Rather, they are merely the most general possible summary of patterns of events within the world. On this view, the planets don't orbit the sun because they "obey" the law of gravity; instead, the law of gravity (or rather, the general theory of relativity, which superseded it) summarizes a regular pattern in nature, a pattern that includes the planetary orbits.

Suppose the laws of physics—even the deepest laws, those that will make up the hoped-for final theory—are indeed just summaries of what goes on in the world. Then how can those laws *explain* anything? Perhaps they can't. That was what Ludwig Wittgenstein thought. "The whole modern conception of the world," Wittgenstein wrote in his *Tractatus*, "is founded on the illusion that the so-called laws of nature are the explanations of natural phenomena. Thus people today stop at the laws of nature, treating them as something inviolable, just as God and Fate were treated in past ages."

Weinberg clearly does not share this Wittgensteinian skepticism. Physicists aren't like priests or oracles. They really do explain things. Explanation is what has happened when they are moved to say, "Aha!" To explain an event scientifically, Weinberg argues, is to show how it fits into the pattern of regularities encoded in some physical principle. And to explain that principle, in turn, is to show that it can be deduced from a more fundamental principle. (Thus, for example, the chemical properties of many molecules can be deduced from—and hence are explained by—the deeper principles of quantum mechanics and electrostatic attraction.) Eventually, according to Weinberg's scheme, all such arrows of scientific explanation will converge at a single bedrock level, the deepest and most comprehensive of all—that of the final theory.

It is conceivable that future physicists will bring the existence of the universe itself into this grand deductive scheme. Perhaps, employing the final theory, they will be able to calculate that the seed of an inflationary multiverse was bound to quantum-tunnel out of nothingness. But what would be the meaning of such a calculation? Would it explain why there is Something rather than Nothing? No. It would merely show that the laws describing the regularities *inside* the world are incompatible with the *nonexistence* of that world. (If, for example, Heisenberg's uncertainty principle says that the value of a field and its rate of change cannot both be precisely zero, then the world as a whole can't very well consist of unchanging nothingness.) To the metaphysical optimist, that might not look like such a bad result. It would mean that the world is, in a sense, self-subsuming, since its existence is entailed, or at least rendered probable, by regularities within it. To the cynic, though, it looks like a vicious circle. Since the world is logically prior to the patterns within it, those internal patterns can't be called on to explain the existence of the world.

My encounter with Weinberg had deepened my understanding of how scientific explanation works. But it had also left me in agreement with him that no such explanation could dispel the mystery of existence. The question *Why is there something rather than nothing?* lies outside the ambit even of the final theory. Notwithstanding the clever imaginative leaps of cosmologists like Stephen Hawking, Ed Tryon, and Alex Vilenkin, a satisfying answer, if there was one, would have to be sought elsewhere, beyond the precincts of theoretical physics.

Would the search prove futile? Perhaps. But that made it all the more noble, in a Sisyphean kind of way. After all, as Weinberg wrote at the very end of *The First Three Minutes*, "The effort to understand the universe is one of the very few things that lifts human life above the level of farce, and gives it some of the grace of tragedy."

Interlude

A Word on Many Worlds

The existence of one world is mysterious enough. But what about the existence of *many* worlds? Such a wanton profusion of being would appear to make the quest for an ultimate explanation all the more hopeless. To the already intractable questions *Why anything?* and *Why this?*, it seems to adjoin a third: *Why so much?*

Yet the hypothesis of many worlds was evidently quite congenial to some of the thinkers I'd encountered. Steven Weinberg, despite his generally skeptical turn of mind, had not been shy about embracing it. Nor had (the rather less skeptical) David Deutsch. Both thought that the existence of multiple universes would render less mysterious certain deep features of our own universe: its otherwise inexplicable quantum behavior (Deutsch), and its improbable suitability for life (Weinberg).

Richard Swinburne, by contrast, had denounced the postulation of "a trillion trillion other universes" as "the height of irrationality." And he is not alone in taking this dim view. The great science-popularizer and fraud-debunker Martin Gardner insisted that "there is not a shred of evidence that there is any universe other than the one we are in." Theories of multiple universes, Gardner said, are "all frivolous fantasies." And the physicist Paul Davies, carrying the debate to the op-ed page of the *New York Times*, declared that "invoking an infinity of unseen universes to explain the unusual features of the one we see is just as ad hoc as invoking an unseen Creator." Each, Davies said, requires a "leap of faith."

Should we or should we not believe in multiple universes? And does our decision have any bearing on the deeper question of why there is Something rather than Nothing?

Before getting to these matters, there's a semantic point to be dealt with. If the universe is "everything there is," then isn't it true by definition that there is only one of these things? Well, yes. But when physicists and philosophers talk about two different regions of spacetime being "two universes," what they generally mean is that those regions are (1) very, very large; (2) causally isolated from each other; and hence (3) mutually unknowable by direct observation. The case for saying the two regions are separate universes is strengthened if (4) they have very different characters—if, for instance, one of them has three spatial dimensions whereas the other has seventeen dimensions. Finally—and here is the existentially titillating possibility—two regions of spacetime might be called separate universes if (5) they are "parallel," meaning that they contain somewhat different versions of the same entities. They might, for instance, contain various alter egos of yourself. Thinkers who entertain the possibility that there are lots of universes in one or another combination of these senses use the term "multiverse" (or sometimes "megaverse") for the entire ensemble of them.

So why believe in the multiverse?

Since other universes are, by definition, not directly observable from our own, the burden of proof is clearly on those who claim they exist. The pro-multiverse camp has essentially two kinds of arguments.

One kind of "pro" argument—the good kind—is that the existence of other universes is entailed by features of our own universe and the theories that best explain those features. For example, measurements of the cosmic background radiation—the echo left over from the Big Bang—indicate that the space we live in is infinite, and that matter is spread randomly throughout it. Therefore, all possible arrangements of matter must exist out there somewhere—including exact and inexact replicas of our own world and the beings in it. A back-of-the-envelope calculation shows that there should be an exact copy of you some 10 to the 10 to the twenty-eighth ($10^{10^{28}}$) meters (or miles or angstroms or light-years—it doesn't really matter what the unit is when the numbers are this large) away. Because of the finite speed of light, however, these parallel worlds—and our doppelgängers within them—are inaccessible to us, and will forever remain so if the expansion of the universe continues to accelerate.

Another, more extravagant kind of multiverse is entailed by the theory of "chaotic inflation." Proposed in the 1980s by the Russian physicist Andrei Linde to explain why our universe looks the way it does—big, uniform, flat, low in entropy—this theory also predicts that Big Bangs should be a fairly routine occurrence. In the inflationary picture, the multiverse is a ferment of innumerable, mutually isolated "bubble universes." These bubble universes do not spring into being out of nothingness; rather, they are held to arise from a sort of preexisting chaos.

The inflationary multiverse thus sheds no light on the mystery of why there is Something rather than Nothing. But it does, as Steven Weinberg observed in our conversation, furnish a neat resolution to another mystery—the mystery of *our* existence. In inflationary cosmology, the laws of nature take the same general form throughout the entire multiverse. However, the details of those laws—the precise strength of the forces, the relative masses of the particles, the number of spatial dimensions, and so on—vary at random from universe to universe. (This randomness is due to quantum fluctuations at the birth of the different bubble universes.) If our own universe is but one among a vast ensemble of universes with randomly varying physical details, it is only to be expected that a few of these universes should have just the right conditions to foster intelligent life. Add to this the truism that, if we exist at all, we are bound to find ourselves in a universe with such life-fostering conditions—the so-called anthropic principle—and the supposed fine-tuning of our universe for life seems wholly unremarkable. No need to invoke the God hypothesis to answer the question *Why are we here?*

So if scientific observations give us reason to think there are other universes, certain mysteries about our own universe go away, almost as a fringe benefit. That was the point that Weinberg had made. But some thinkers want to turn this reasoning around. They insist that other universes *must* exist precisely to make certain mysteries go away. This is the second kind of argument for the multiverse—the bad kind, since it has nothing to do with empirical observation.

One version of this argument derives from attempts to make sense of quantum theory. Take the famous paradox of Schrödinger's cat—the unfortunate feline in a box who, because of the quantum superposition of possibilities, is simultaneously alive and dead. According to the "many-worlds" interpretation of quantum theory, Schrödinger's thought experiment splits the universe into two parallel copies, one with a live cat, the other with a dead one (and each of them with a version of you). Physicists who look favorably on this interpretation—and many distinguished ones have, among them Richard Feynman, Murray Gell-Mann, and Stephen Hawking—claim that each universe splits into copies every second numbering something like 10 followed by one hundred zeros, all of them equally real. Yet, since quantum theory forbids these parallel worlds from interacting in any but the ghostliest of ways, their reality cannot be experimentally observed.

Another version of this backward argument for multiple universes was championed by the late Princeton philosopher David K. Lewis. Lewis shocked his fellow philosophers by claiming that all logically possible worlds are real—just as real as the one we call the "actual" world. Why did he think such a thing? Because, he said, their reality would crisply solve a wide range of philosophical problems. Take the problem of counterfactuals. What does it mean to say, "If JFK hadn't gone to Dallas, the Vietnam War would have ended earlier"? According to Lewis, the counterfactual statement is true only if there is a possible world very similar to the actual world in which JFK *didn't* go to Dallas and the Vietnam War *did* end earlier. Lewis's possible worlds are also useful for making sense of propositions beginning, "If pigs could fly . . ."

Such dubious arguments in support of the multiverse idea have evoked equally dubious arguments against it—such as these three:

(1) *It's not science.* Both Paul Davies and Martin Gardner submit that the proposition "the multiverse exists" has no empirical content, and hence

amounts to empty metaphysics. But some of the theories that *imply* the existence of a multiverse—like the theory of chaotic inflation—*do* lead to testable predictions; moreover, these predictions have been borne out by the evidence collected so far. And improved measurements over the next decade of the microwave background radiation and of the large-scale distribution of matter may further confirm these theories—or overturn them. That looks like real science.

(2) *Alternative universes should be shaved away by Occam's razor.* Both Davies and Gardner complain that the multiverse notion is too extravagant. "Surely the conjecture that there is just one universe and its Creator is infinitely simpler and easier to believe than that there are countless billions upon billions of worlds," Gardner writes. Is it? Our universe came into being with the Big Bang, and (as the Canadian philosopher John Leslie has observed) it would be exceedingly odd if the mechanism behind this world-engendering event bore the label "THIS MECHANISM OPERATED ONLY ONCE." A computer program that prints out the entire sequence of numbers is simpler than one that prints out only a single, very long number.

(3) *The multiverse, if real, would reduce our own world to a* Matrix-*like simulation.* This objection, voiced by Davies, is surely the most bizarre of the lot. If there really were myriad universes, Davies argues, then some of them would contain advanced technological civilizations that could use computers to simulate endless *virtual* worlds. These virtual worlds would vastly outnumber the actual universes that made up the multiverse. So, he continues, taking the multiverse theory at face value, it is much more likely that we ourselves are creatures in a virtual world than in an actual physical universe. If the multiverse theory is true, Davies says, "there is no reason to expect our world—the one in which you are reading this right now—to be real as opposed to a simulation." He takes this to be a *reductio ad absurdum* of the multiverse idea. But Davies's argument is a poor one, for at least two reasons. If it were valid, it would rule out the existence of technologically advanced civilizations in *this* universe, since they too would presumably create simulated worlds in great abundance. And the hypothesis that we are living in a simulation itself has no empirical content. How could it be verified or falsified? We cannot even talk about it coherently, as Hilary

Putnam has pointed out, since our words could refer only to things "inside" the alleged simulation.

Among those who take the idea of the multiverse seriously, perhaps the greatest disagreement is over how many distinct versions of it there are. Is, for instance, the "quantum multiverse" the same as the "inflationary multiverse"? The quantum multiverse, as I mentioned earlier, is the version that is invoked to make sense of quantum weirdness. First put forth in the 1950s by the physicist Hugh Everett III in the form of his many-worlds interpretation, it says that the different possible outcomes of a quantum measurement correspond to parallel universes, all coexisting in some sort of larger reality. The inflationary multiverse, by contrast, was suggested by cosmological considerations. It encompasses an infinity of bubble universes, each arising with its own Big Bang out of a primordial chaos.

The worlds making up the inflationary multiverse are separated from one another by regions of space that, since they are expanding faster than the speed of light, cannot be traversed. By contrast, the worlds making up the quantum multiverse are separated from one another by . . . well, no one can quite say. The image of quantum worlds "branching off" from one another suggests that they are in some sense close together; so does the notion of such parallel worlds jostling one another ever so slightly (as in the double-slit experiment).

Given such differences, one might think that we are talking about two distinct species of multiverse here. Surprisingly, though, there are distinguished physicists who happily conflate the two. One such is Leonard Susskind, a coinventor of string theory. "The many-worlds [multiverse] of Everett seems, at first sight, to be quite a different conception than the eternally-inflating universe," Susskind has observed. "However, I think the two may really be the same thing."

Susskind's belief in the identity of these two seemingly distinct versions of the multiverse puzzled me, so I made sure to mention it to Steven Weinberg. "I found it puzzling too," he said. "I've spoken to other people about it, and they don't understand it either." Although Weinberg is himself sympathetic to the many-worlds interpretation of quantum mechanics, he finds it "completely perpendicular" to the issue of the inflationary multiverse. In other words, Weinberg could see no reason to equate the two multiverses, as Susskind did. "I don't agree with Susskind on that," Weinberg told me, "and I don't know why he said it."

Whether the multiverses posited by physicists are one or many, they are certainly *contingent*, not necessary. There is nothing within them that explains *why* they exist. And the individual worlds a multiverse comprises, while randomly varying in their features, nevertheless obey the same laws of nature—laws that inexplicably take one particular form rather than another. So even the most extravagant multiverse, conceived in merely physical terms, leaves unresolved a pair of fundamental questions: Why *these* laws? And why should there be a multiverse that embodies them, rather than nothing at all?

"It is probable that there is some secret here which remains to be discovered," observed the great nineteenth-century American pragmatist philosopher C. S. Peirce—the same thinker, as it happens, who mockingly regretted that universes are not "as plentiful as blackberries." Physics alone seems impotent to discover this secret. And that has driven some physicists to flirt with—and even embrace—a mystical way of thinking about reality which harkens back to Plato, if not to Pythagoras.

10

PLATONIC REFLECTIONS

See *Mystery* to *Mathematics* fly!
In vain! they gaze, turn giddy, rave, and die.
—ALEXANDER POPE, The Dunciad

Mysticism and mathematics go way back together. It was the mystical cult of Pythagoras that, in ancient times, invented mathematics as a deductive science. "All is number," proclaimed Pythagoras—by which he seemed to mean that the world was quite literally constituted by mathematics. It is little wonder that the Pythagoreans worshipped numbers as a divine gift. (They also believed in the transmigration of souls and held the eating of beans to be wicked.)

Today, two and a half millennia later, mathematics still has a whiff of the mystical about it. A majority of contemporary mathematicians (a typical, though disputed, estimate is about two-thirds) believe in a kind of heaven—not a heaven of angels and saints, but one inhabited by the perfect and timeless objects they study: n-dimensional spheres, infinite numbers, the square root of -1, and the like. Moreover, they believe that they commune with this realm of timeless entities through a sort of extrasensory perception. Mathematicians who buy into this fantasy are called "Platonists," since their mathematical heaven resembles the transcendent realm described by Plato in his *Republic*. Geometers, Plato observed, talk about circles that are perfectly round and infinite lines that are perfectly straight. Yet such perfect entities are nowhere to be found in the world we perceive with our senses. The same is true, Plato held, of numbers. The number 2, for instance, must be composed of a pair of perfectly equal units; but no two things in the sensible world are perfectly equal.

Plato concluded that the objects contemplated by mathematicians must exist in another world, one that is eternal and transcendent. And today's mathematical Platonists agree. Among the most distinguished of them is Alain Connes, holder of the Chair of Analysis and Geometry at the Collège de France, who has averred that "there exists, independently of the human mind, a raw and immutable mathematical reality." Another contemporary Platonist is René Thom, who became famous in the 1970s as the father of catastrophe theory. "Mathematicians should have the courage of their most profound convictions," Thom has declared, "and thus affirm that mathematical forms indeed have an existence that is independent of the mind considering them."

Platonism is understandably seductive to mathematicians. It means that the entities they study are no mere artifacts of the human mind: these entities are discovered, not invented. Mathematicians are like seers, peering out at a Platonic cosmos of abstract forms that is invisible to lesser mortals. As the great logician Kurt Gödel, among the staunchest of Platonists, put it, "We do have something like a perception" of mathematical objects, "despite their remoteness from sense experience." And Gödel was quite sure that the Platonic realm which mathematicians imagined themselves to be perceiving was no collective hallucination. "I don't see any reason why we should have less confidence in this kind of perception, i.e., in mathematical intuition, than in sense perception," he declared. (Gödel also believed in the existence of ghosts, but that is another matter.)

Many physicists also feel the allure of Plato's vision. Not only do mathematical entities seem to be "out there"—eternal, objective, immutable—they also appear to be sovereign over the physical universe. How else can we account for what the physicist Eugene Wigner famously called the "unreasonable effectiveness of mathematics in the natural sciences"? Mathematical beauty has time and again proved to be a reliable guide to physical truth, even in the absence of empirical evidence. "You can recognize truth by its beauty and simplicity," said Richard Feynman. "When you get it right, it is obvious that it's right." If, in Galileo's phrase, the "book of nature is written in the language of mathematics," this could only be because the natural world itself is inherently mathematical. As the astronomer James Jeans picturesquely put it, "God is a mathematician."

To a devout Platonist, though, this invocation of God is merely a bit of superfluous poetry. Who needs a creator when mathematics by itself might be capable of engendering and sustaining a universe? Mathematics feels real, and the world feels mathematical. Could it be that mathematics furnishes the key to the mystery of being? If mathematical entities do exist, as the Platonists believe, they must exist necessarily, from all eternity. Perhaps this eternal mathematical richness somehow spilled over into a physical cosmos—a cosmos of such complexity that it gave rise to conscious beings who are able to make contact with the Platonic world whence they ultimately sprang.

This is a pretty picture. But could anyone who is not in the habit of eating lotus leaves take it seriously?

I had the impression that at least one quite rigorous thinker did: Sir Roger Penrose, the emeritus Rouse Ball Professor of Mathematics at Oxford. Penrose is among the most formidable mathematical physicists alive. He has been hailed by fellow physicists, notably Kip Thorne, for bringing higher mathematics back into theoretical physics after a long period in which the two fields had ceased to communicate. In the 1960s, working with Stephen Hawking, Penrose used sophisticated mathematical techniques to prove that the expansion of the universe out of the Big Bang must have been a precise reversal of the collapse of a star into a black hole. In other words, the universe must have begun as a singularity. In the 1970s, Penrose developed the "cosmic censorship hypothesis," which says that every singularity is cloaked by an "event horizon" that protects the rest of the universe from the breakdown of physical laws. Penrose has also been a pioneer of "twistor theory," a beautiful new approach to reconciling general relativity with quantum mechanics. In 1994, Queen Elizabeth bestowed a knighthood on Penrose for such achievements.

Penrose also has a penchant for oddities. As a graduate student, he became obsessed with "impossible objects"—that is, physical objects that seem to defy the logic of three-dimensional space. His success in "constructing" one such impossible object, now known as the "Penrose tribar," inspired M. C. Escher to create two of his most famous prints: *Ascending and Descending*, which depicts a gaggle of monks endlessly tromping up—or down?—a staircase that goes nowhere, and *Waterfall*, which shows a perpetually descending circuit of water. (I once heard the philosopher Arthur Danto say that every philosophy department should keep an impossible object around the office, to instill a sense of metaphysical humility.)

Penrose, I knew, is an unabashed Platonist. Over the years, in his writings and public lectures, he has made it clear that he takes mathematical entities to be as real and mind-independent as Mount Everest. Nor has he been shy about invoking the name of Plato himself. "I imagine that whenever the mind perceives a mathematical idea it makes contact with Plato's world of mathematical concepts," he wrote in his 1989 book, *The Emperor's New Mind*. "The mental images that each [mathematician] has, when making this Platonic contact, might be rather different in each case, but communication is possible because each is directly in contact with the *same* eternally existing Platonic world!"

What really piqued my interest, though, was Penrose's occasional hint that our own world was an outcropping of this Platonic world. I first noticed such hints in his second book for a popular audience, *Shadows of the Mind*, which came out in 1994 and, like its intellectually daunting predecessor, was an improbable bestseller. Penrose began by arguing, based on an appeal to Gödel's incompleteness theorem, that the human mind had powers of mathematical discovery that exceeded those of any possible computer. Such powers, Penrose contended, must be essentially quantum in nature. And they would be understood only when physicists had discovered a theory of quantum gravity—the holy grail of contemporary physics. Such a theory would finally make sense of the bizarre interface between the quantum world and classical reality—and, in the bargain, it would reveal how the human brain leapfrogs the bounds of mechanical computation into full Technicolor consciousness.

Penrose's ideas on consciousness did not impress many brain scientists. As the late Francis Crick irritably jibed, "His argument is that quantum gravity is mysterious and consciousness is mysterious and wouldn't it be wonderful if one explained the other." Yet there was more to Penrose's agenda than that. The very title of his book, *Shadows of the Mind*, was a double entendre. On the one hand, it was meant to suggest that the electrical activities of our brain cells, usually thought to be the cause of our mental life, are mere "shadows" cast by deeper quantum processes going on in the brain, which are the true springs of consciousness.

On the other hand, "shadows" harkened back to Plato—specifically, to Plato's Allegory of the Cave, from Book VII of the *Republic*. In this allegory, Plato likens us to prisoners chained in a cave and condemned to look only at the rock wall in front of them. On that wall they see a play of shadows, which they take for reality. Little do they realize that there is a world of real things behind them which is the source of these shadowy images. If one of the prisoners were to be liberated from the cave, he would initially be blinded by the sunlight outside. But as his eyes adjusted, he would come to understand his new surroundings. And what would happen if he returned to the cave to tell his fellow prisoners about the real world? Unused to the darkness after his time in the sunlight, he would be unable to make out the shadows they took for reality. His tale of a real world outside the cave would "provoke laughter." The other prisoners would say that "he had returned from his journey aloft with his eyes ruined," and that "it was not worthwhile even to attempt the ascent."

This outside world in the Allegory of the Cave stands for the timeless realm of Forms, wherein genuine reality resides. For Plato, the inhabitants of this realm included abstractions like Goodness and Beauty, as well as the perfect objects of mathematics. Was Penrose, in suggesting that what we took to be reality consisted of "shadows" of such a realm, merely trafficking in neo-Platonist mysticism? Or did his almost unrivalled grasp of quantum theory and relativity, of singularities and black holes, of higher mathematics and the nature of consciousness, afford him genuine insight into the mystery of existence?

I did not have to journey far to obtain enlightenment on this matter. Waiting for the elevator one day in the lobby of the mathematics building at New York University, I saw an announcement that Penrose would soon be coming to Manhattan. He had been invited to deliver an endowed series of lectures on his contributions to theoretical physics. I went home and called his publicist at Oxford University Press to see if an interview could be arranged. A couple of days later, she rang back to say that "Sir Roger" had agreed to make a little time for me to talk about philosophy.

As it happened, Penrose was being put up in a stately apartment building fronting the west side of Washington Square, just steps away from my own Greenwich Village residence. On the appointed day, I set out across the square, which, given the glorious spring weather, was a typically cacophonous buzz of activity. Here, a pick-up jazz combo played for people lounging in the grass; there, a would-be Bob Dylan was wailing away over his guitar. By the big fountain in the middle of the square, banjee boys did improvised gymnastics for an earnest-looking audience of European tourists, while the dogs in the nearby dog run capered and barked.

I exited the square at the northwest corner, where the chess hustlers congregate over the outdoor chess tables, waiting for naive passersby to join them in a game and lose a little money. Glancing up at the old Earle Hotel near the corner, I recalled reading somewhere that it was during a stay in that very hotel many decades ago that the Mamas and the Papas had written their hit song "California Dreamin'." Inevitably, that was the tune running through my head as I entered the lobby of the building where Penrose was staying, which was vaguely Moorish in decor. The liveried doorman told me to take the elevator to the penthouse.

Sir Roger himself answered the door. He was an elfin man who, with his thick head of auburn hair, looked much younger than his years. (He was born in 1931.) The apartment where he was staying was quite grand, in a prewar New York sort of way. Its high ceilings were embellished with elaborate moldings, and great casement windows with leaden mullions overlooked the treetops of Washington Square. By way of small talk, I pointed out an enormous elm, reputed to be the oldest tree in Manhattan, and told Sir Roger that it was known as the "hanging tree," since it had been used for executions in the late eighteenth century. He nodded genially at this bit of unsolicited information, then padded off to the kitchen to fetch me a cup of coffee.

Why, I briefly wondered as I took a seat on the sofa, did everyone but me seem to find caffeinated beverages more conducive than alcohol to pondering the mystery of existence?

When Sir Roger returned, I asked him whether he really believed in a Platonic world, one that exists over and above the physical world. Wouldn't such a two-world view be a little profligate, ontologically speaking?

"Actually, there are *three* worlds," he replied, warming to my challenge. "Three worlds! And they're all separate from one another. There's the Platonic world, there's the physical world, and there's also the mental world, the world of our conscious perceptions. And the interconnections among these three worlds are *mysterious*. The main mystery I've been addressing, I suppose, is how the mental world is related to the physical world: how certain types of highly organized physical objects—our brains—seem to produce conscious awareness. But another mystery—which, to a mathematical physicist, is just as deep—is the relationship between the Platonic world and the physical world. As we search for the deepest possible understanding of how the world behaves, we're driven to mathematics. It's almost as though the physical world is *built* out of mathematics!"

So he was more than a Platonist—he was a Pythagorean! Or he was at least flirting with Pythagoras's mystical doctrine that the world was constituted by mathematics: *all is number.* Yet I noticed that there was one link among his three worlds that Penrose hadn't yet addressed. He had mentioned how the mental world might be linked to the physical world, and how the physical world might be linked to the Platonic world of abstract mathematical ideas. But what about the supposed link between that Platonic world and the mental world? How are our minds supposed to get in touch with these incorporeal Platonic Forms? If we are to have knowledge of mathematical entities, we have to "perceive" them in some way, as Gödel put it. And perceiving an object usually means having causal commerce with it. I perceive the cat on the mat, for example, because photons emitted from the cat are impinging on the retinas of my eyes. But Platonic Forms aren't like the cat on the mat. They don't live in the world of space and time. There are no photons traveling back and forth between them and us. So we can't perceive them. And if we can't perceive mathematical objects, how can we possibly come to have knowledge of them?

Plato believed that such knowledge was derived from a previous existence, before we were born, during which our souls communed directly
with the Forms; what we know of mathematics—and of Beauty and Goodness, for that matter—thus consists of "reminiscences" from this disem-
bodied existence that preceded our earthly lives. Nobody takes that idea
seriously any more. Yet what's the alternative? Penrose himself had written that human consciousness somehow "breaks through" to the Platonic
world when we contemplate mathematical objects. But consciousness
depends on physical processes in the brain, and it's hard to see how such
processes could be affected by a nonphysical reality.

When I put this objection to Penrose, he furrowed his brow and was
silent for a moment. "I know that's something philosophers worry about,"
he said at length. "But I'm not sure I've ever really understood the argument. It's *out there*, the Platonic world, and we *can* have access to it. Ultimately, our physical brains are constructed out of material that is itself
intimately related to the Platonic world of mathematics."

So was he saying that we can perceive mathematical reality because our
brains are somehow a *part* of that reality?

"It's a little more complicated than that," Sir Roger corrected me. "Each
of the three worlds—the physical world, the world of consciousness, and
the Platonic world—emerges out of a tiny bit of one of the others. And
it's always the most *perfect* bit. Consider the human brain. If you look
at the entire physical cosmos, our brains are a tiny, tiny part of it. But
they're the most perfectly organized part. Compared to the complexity
of a brain, a galaxy is just an inert lump. The brain is the most exquisite
bit of physical reality, and it's just this bit that gives rise to the mental
world, the world of conscious thought. Likewise, it's only a small part of
our conscious thought that connects us to the Platonic world, but it's the
purest part—the part that consists in our contemplation of mathematical
truth. Finally, just a few bits of the mathematics in the Platonic world are
needed to describe the entire physical world—but they're quite the most
powerful and extraordinary bits!"

Spoken like a true mathematical physicist, I thought to myself. But could
it be that these "powerful and extraordinary" bits of mathematics—the bits

that preoccupy Penrose—are so powerful that they can generate a physical world all by themselves? Does mathematics carry its own ontological clout?

"Something like that, yes," Sir Roger said. "Maybe philosophers worry too much about lesser issues without realizing that this is perhaps the greatest mystery of all: how the Platonic world 'controls' the physical world."

He paused for a moment to reflect, and then added, "I'm not saying I can *resolve* this mystery."

After a bit of small talk about Gödel's incompleteness theorems, quantum computation, artificial intelligence, and animal consciousness ("I have no idea whether a starfish is conscious," Penrose said, "but there *should* be some observable signs"), my visit with Sir Roger came to an end. I left his penthouse world of Platonic ideas and, after a quick elevator descent, reentered the ephemeral world of sensuous appearances below. Retracing my steps through Washington Square, I walked under the "hanging tree," past the chess hustlers, and into the crowded plaza around the central fountain, encountering the same chaos of exuberant motion, garish colors, pungent smells, and exotic noises. These people! I thought, What do they know of the serene and timeless Platonic realm? Whether they be tourists or buskers, panhandlers or adolescent anarchists, or even NYU professors of cultural studies taking a shortcut through the square on their way to a lecture, their consciousness never touches the ethereal realm of mathematical abstraction that is the true source of reality. Little did they realize that, despite the abundant sunshine, they were chained in the allegorical darkness of Plato's cave, condemned to live in a world of shadows. They could have no genuine knowledge of reality. That was open only to those who apprehended the eternal Forms, the true philosophers—like Penrose.

But gradually the spell that Sir Roger had cast on me began to wear off. How could the solemn mathematical abstractions in Plato's heaven have given rise to the gaiety of life in Washington Square? Do such abstractions really hold the answer to the mystery of why there is Something rather than Nothing?

The scheme of being that Penrose had conjured up for me seemed almost miraculously self-creating and self-sustaining. There are three worlds: the Platonic world, the physical world, and the mental world. And each of the worlds somehow engenders one of the others. The Platonic world, through the magic of mathematics, engenders the physical world. The physical world, through the magic of brain chemistry, engenders the mental world. And the mental world, through the magic of conscious intuition, engenders the Platonic world—which, in turn, engenders the physical world, which engenders the mental world, and so on, around and around. Through this self-contained causal loop—Math creates Matter, Matter creates Mind, and Mind creates Math—the three worlds mutually support one another, hovering in midair over the abyss of Nothingness, like one of Penrose's impossible objects.

Yet, despite what this picture might suggest, the three worlds are not ontologically coequal. It is the Platonic world, in Penrose's vision, that is the *fons et origo* of reality. "To me the world of perfect forms is primary, its existence being almost a logical necessity—and *both* the other worlds are its shadows," he wrote in *Shadows of the Mind*. The Platonic world, in other words, is compelled to exist by logic alone, and the contingent world—the world of matter and mind—follows as a shadowy by-product. That's Penrose's solution to the puzzle of existence.

And it left me with two misgivings. Is the existence of the Platonic world really assured by logic itself? And even if it is, what then makes it cast shadows?

As to the first, I couldn't help noticing what looked like a failure of nerve on Penrose's part. The existence of the Platonic world, he said, is "almost a logical necessity." Why this "almost"? Logical necessity is not a thing that admits of degree. It is all or nothing. Penrose makes much of the alleged fact that the Platonic world of mathematics is "eternally existing," that its reality is "profound and timeless." But the same, one might note, would be true of God—*if* God existed. Yet God is not a logically necessary being; his existence can be denied without contradiction. Why should mathematical objects be superior to God in this respect?

The belief that the objects of pure mathematics exist necessarily has been called an "ancient and honorable" one, but it doesn't hold up terribly well under scrutiny. It seems to be based on two premises: (1) mathematical truths are logically necessary; and (2) some of those truths assert the existence of abstract objects. As an example, consider proposition twenty in Euclid's *Elements*, which says that there are infinitely many prime numbers. This certainly looks like an existence claim. Moreover, it appears to be true as a matter of logic. Indeed, Euclid proved that denying the existence of an infinity of primes led straight to absurdity. Suppose there were only finitely many prime numbers. Then, by multiplying them all together and adding 1, you would get a new number that was bigger than all the primes and yet divisible by none of them—contradiction!

Euclid's *reductio ad absurdum* proof of the infinity of prime numbers has been called the first truly elegant bit of reasoning in the history of mathematics. But does it give any grounds for believing in the existence of numbers as eternal Platonic entities? Not really. In fact, the existence of numbers is *presupposed* by the proof. What Euclid really showed was that *if* there are infinitely many things that behave like the numbers 1, 2, 3, . . . , *then* there must be infinitely many things among them that behave like *prime* numbers. All of mathematics can be seen to consist of such *if-then* propositions: *if* such-and-such a structure satisfies certain conditions, *then* that structure must satisfy certain further conditions. These *if-then* truths are indeed logically necessary. But they do not entail the existence of any objects, whether abstract or material. The proposition "2 + 2 = 4," for example, tells you that *if* you had two unicorns and you added two more unicorns, *then* you would end up with four unicorns. But this *if-then* proposition is true even in a world that is devoid of unicorns—or, indeed, in a world that contains nothing at all.

Mathematicians essentially make up complex fictions. Some of these fictions have analogues in the physical world; they compose what we call "applied mathematics." Others, like those positing higher infinities, are purely hypothetical. Mathematicians, in creating their imaginary universes, are constrained only by the need to be logically consistent—and to create something of beauty. (" 'Imaginary universes' are so much more beautiful than the stupidly constructed 'real' one," declared the great English number theorist G. H. Hardy.) As long as a collection of axioms does not lead to a contradiction, then it is at least *possible* that it describes something. That is why, in the words of Georg Cantor, who pioneered the theory of infinity, "the essence of mathematics is freedom."

So the existence of mathematical objects is not mandated by logic, as Penrose seemed to believe. It is merely *permitted* by logic—a much weaker conclusion. Practically anything, after all, is permitted by logic. But for some modern-day Platonists of an even more radical stripe, that seems to be permission enough. As far as they are concerned, self-consistency alone guarantees mathematical existence. That is, as long as a set of axioms does not lead to a contradiction, then the world it describes is not only possible—it is actual.

One such radical Platonist is Max Tegmark, a young Swedish-American cosmologist who teaches at MIT. Tegmark believes, like Penrose, that the universe is inherently mathematical. Also like Penrose, he believes that mathematical entities are abstract and immutable. Where he goes beyond Sir Roger is in holding that *every* consistently describable mathematical structure exists in a genuine physical sense. Each of these abstract structures constitutes a parallel world, and together these parallel worlds make up a mathematical multiverse. "The elements of this multiverse do not reside in the same space but exist outside of space and time," Tegmark has written. They can be thought of as "static sculptures that represent the mathematical structure of the physical laws that govern them."

Tegmark's extreme Platonism furnishes a very cheap resolution to the mystery of existence. It is basically, as he concedes, a mathematical version of Robert Nozick's principle of fecundity, which says that reality encompasses all logical possibilities, that it is as rich and variegated as it can be. Anything that is possible must actually exist—hence the triumph of Something over Nothing. What makes such a principle compelling for Tegmark is the peculiar ontological muscle that mathematics seems to possess. Mathematical structures, he says, "have an eerily real feel to them." They are fruitful in uncovenanted ways; they surprise us; they "bite back." We get more out of them than we seem to have put into them. And if something *feels* so real, it must *be* real.

But why should we be swayed by this "real feel," no matter how eerie?

Tegmark and Penrose may be swayed, but another great physicist, Richard Feynman, was decidedly not. "It's just a feeling," Feynman once said dismissively, when asked whether the objects of mathematics had an independent existence.

Bertrand Russell came to take an even sterner view of such mathematical romanticism. In 1907, when he was in his relatively youthful thirties, Russell penned a gushing tribute to the transcendent glories of mathematics. "Rightly viewed," he wrote, mathematics "possesses not only truth, but supreme beauty—a beauty cold and austere, like that of sculpture." Yet by his late eighties he had come to view his callow rhapsodizing as "largely nonsense." Mathematics, the aged Russell wrote, "has ceased to seem to me non-human in its subject-matter. I have come to believe, though very reluctantly, that it consists of tautologies. I fear that, to a mind of sufficient intellectual power, the whole of mathematics would appear trivial, as trivial as the statement that a four-footed animal is an animal."

How can the romantic Platonism of Penrose, Tegmark, and others survive Russell's cold cynicism? Well, if neither logic nor feeling can underwrite the existence of timeless mathematical Forms, then perhaps science can. Our best scientific theories of the world, after all, incorporate quite a lot of mathematics. Take Einstein's general theory of relativity. In describing how the shape of spacetime is determined by the way matter and energy are distributed throughout the universe, Einstein's theory invokes a host of mathematical entities, like "functions," "manifolds," and "tensors." If we believe that the theory of relativity is true, then aren't we committed to the existence of these entities? Isn't it intellectually dishonest to pretend they aren't real if they are indispensable to our scientific understanding of the world?

That, in a nutshell, is the so-called Indispensability Argument for mathematical existence. It was originally proposed by Willard Van Orman Quine, the dean of twentieth-century American philosophy and the man who famously declared, "To be is to be the value of a variable." Quine was the ultimate "naturalist" philosopher. For him, science was the final arbiter of existence. And if science inescapably refers to mathematical abstractions, then those abstractions *exist*. Although we don't observe them directly, we need them to explain what we do observe. As one philosopher put it, "We have the same kind of reason for believing in numbers and some other mathematical objects as we have for believing in dinosaurs and dark matter."

The Indispensability Argument has been called the only argument for mathematical existence that is worth taking seriously. But even if it is valid, it provides scant comfort for Platonists like Penrose and Tegmark. It robs mathematical Forms of their transcendence. They become mere theoretical posits that help explain our observations. They are on par with physical entities like subatomic particles, since they occur in the same explanations. How can they be responsible for the *existence* of the physical world if they themselves are part of the very *fabric* of that world?

And it gets worse for the Platonists. Mathematics, it turns out, may not be indispensable to science after all. It may be that we can explain how the physical world works without invoking abstract mathematical entities, just as we have learned to do so without invoking God.

One of the first to raise this possibility was the American philosopher Hartry Field. In his 1980 book, *Science without Numbers*, Field showed how Newton's theory of gravitation—which, on the face of it, is mathematical through and through—could be reformulated so that it made no reference whatsoever to mathematical entities. Yet the numbers-free version of Newton's theory would yield exactly the same predictions, though in a rather more roundabout way.

If the program of "nominalizing" science—that is, of stripping away its mathematical trappings—could be extended to theories like quantum mechanics and relativity, it would mean that Quine was wrong. Mathematics is not "indispensable." Its abstractions need play no role in our understanding of the physical world. They are just a glorified accounting device—nice in practice (since they lead to shorter derivations), but dispensable in theory. To creatures of greater intelligence elsewhere in the cosmos, they might not be necessary at all. Far from being timeless and transcendent, numbers and other mathematical abstractions would be exposed as mere terrestrial artifacts. We could banish them from our ontology the way the protagonist of Bertrand Russell's story "The

Mathematician's Nightmare" did—with a cry of "Avaunt! You are only Symbolic Conveniences!"

But would that spell the doom of Platonism as a resolution to the mystery of existence? Maybe not. Recall that there was something missing from Roger Penrose's Platonic scheme. The worlds of matter and consciousness were "shadows," he held, of the Platonic world of mathematics. But what, in this metaphor, was the source of the illumination that allowed the Forms to cast their shadows? Sir Roger conceded that it was a "mystery" how mathematical abstractions could be creatively effective. Such abstractions are supposed to be causally inert: they neither sow nor reap. How could mere passive patterns, however perfect and timeless, reach out and make a world?

Plato himself had no such lacuna in his scheme. For him there *was* a source of light, a metaphorical Sun. And that was the Form of the Good. Goodness, in Plato's metaphysics, stands above the lesser Forms, including the mathematical ones. Indeed, it stands above the Form of Being: "the Good is itself not existence, but far beyond existence in dignity," as Socrates tells us in Book VI of Plato's *Republic*. It is the Form of the Good that "bestows existence upon things"—not by free choice, the way the Christian God is supposed to have done, but by logical necessity. Goodness is the ontological Sun. It shines beams of Being on the lesser Forms, and they in turn cast a shadowy play of Becoming—which is the world we live in.

So that is Plato's vision of the Good as a sunlike source of reality. Should we dismiss it as a woolly poetical conceit? It seems even less helpful than Penrose's own mathematical Platonism at resolving the mystery of existence. Who could imagine that abstract Goodness might bear creative responsibility for a cosmos like ours, which is un-good in so many ways? Yet I was surprised to find that there was at least one thinker who did imagine precisely such a thing. And I was still more surprised to discover that he had managed to convince some of the world's leading philosophers that he might not be entirely daft in doing so. Yet somehow I wasn't surprised to learn that he lived in Canada.

Interlude

It from Bit?

Mathematical Platonism turned out to be a nonstarter as an ultimate explanation of being. But its shortcomings invite deeper reflection on the nature of reality.

Of what does reality, at the most fundamental level, consist? It was Aristotle who supplied the classic answer to this question:

Reality = Stuff + Structure

This Aristotelian doctrine is known as "hylomorphism," from the Greek *hyle* (stuff) and *morphe* (form, structure). It says that nothing really exists unless it is a composite of structure and stuff. Stuff without structure is chaos—tantamount, in the ancient Greek imagination, to nothingness. And structure without stuff is the mere ghost of being, as ontologically wispy as the smile of the Cheshire Cat.

Or is it?

Over the last few centuries, science has relentlessly undermined this Aristotelian understanding of reality. The better our scientific explanations get, the more that "stuff" tends to drop out of the picture. The dematerialization of nature began with Isaac Newton, whose theory of gravity invoked the seemingly occult notion of "action at a distance." In Newton's system, the Sun reached out and exerted its gravitational pull on Earth, even though there was nothing but empty space between them. Whatever the mechanism of influence between the two bodies might be, it seemed to involve no intervening "stuff." (Newton himself was coy on how this could be, declaring *Hypotheses non fingo*—"I frame no hypotheses.")

If Newton dematerialized nature on the largest of scales, from the solar system on up, modern physics has done the same on the smallest of scales, from the atom on down. In 1844, Michael Faraday, observing that matter could be recognized only by the forces acting on it, asked, "What reason is there to suppose that it exists at all?" Physical reality, Faraday proposed, actually consists not of matter but of *fields*—that is, of purely mathematical structures defined by points and numbers. In the early twentieth century, atoms, long held up as paragons of solidity, were discovered to be mostly empty space. And quantum theory revealed that their subatomic constituents—electrons, protons, and neutrons—behaved more like bundles of abstract properties than like little billiard balls. At each deeper level of explanation, what was thought to be stuff has given way to pure structure. The latest development in this centuries-long trend toward the dematerialization of nature is string theory, which builds matter out of pure geometry.

The very notion of *impenetrability*, so basic to our everyday understanding of the material world, turns out to be something of a mathematical illusion. Why don't we fall through the floor? Why did the rock rebound when Dr. Johnson kicked it? Because two solids can't interpenetrate each other, that's why. But the reason they can't has nothing to do with any sort of intrinsic stufflike solidity. Rather, it's a matter of numbers. To squash two atoms together, you'd have to put the electrons in those atoms into numerically identical quantum states. And that is forbidden by something in quantum theory called the "Pauli exclusion principle," which allows two electrons to sit directly on top of each other only if they have opposite spins.

As for the sturdiness of individual atoms, that too is essentially mathematical. What keeps the electrons in an atom from collapsing into the nucleus? Well, if the electrons were sitting right on top of the nucleus, we'd know exactly where each electron was (right in the center of the atom) and how fast it was moving (not at all). And that would violate Heisenberg's uncertainty principle, which does not permit the simultaneous determination of a particle's position and momentum.

So the solidity of the ordinary material objects that surround us—tables and chairs and rocks and so forth—is a joint consequence of the Pauli exclusion principle and Heisenberg's uncertainty principle. In other words, it comes down to a pair of abstract mathematical relations. As the poet Richard Wilbur wrote, "Kick at the rock, Sam Johnson, break your bones: / But cloudy, cloudy is the stuff of stones."

At its most fundamental, science describes the elements of reality in terms of their relations to one another, ignoring any stufflike quiddity those elements might possess. It tells us, for example, that an electron has a certain mass and charge, but these are mere propensities for the electron to be acted upon in certain ways by other particles and forces. It tells us that mass is equivalent to energy, but it gives us no idea of what energy really *is*—beyond being a numerical quantity that, when calculated correctly, is conserved in all physical processes. As Bertrand Russell noted in his 1927 book, *The Analysis of Matter*, when it comes to the intrinsic nature of the entities making up the world, science is silent. What it presents us with is one great relational web: all structure, no stuff. The entities making up the physical world are like the pieces in a game of chess: what counts is the role defined for each piece by a system of rules that say how it can move, not the stuff that the piece is made of.

The physicist's view of reality, by the way, is remarkably akin to the view of language proposed over a century ago by Ferdinand de Saussure, the father of modern linguistics. Language, Saussure maintained, is a purely relational system. Words have no inner essence. The intrinsic character of the noises we make when talking is irrelevant to communication; the important thing is the system of *contrasts* among those noises. This is what Saussure meant when he wrote that "in language, there are only differences *without positive terms*." Saussure's elevation of structure over stuff was the inspiration for the structuralist movement that swept aside existentialism in France in the late 1950s. It was taken up in anthropology by

Claude Lévi-Strauss and in literary theory by Roland Barthes. Its extension to the universe as a whole might well be called "cosmic structuralism."

If reality were indeed pure structure, that would open up radically new ways of thinking about it. One of these is the way of Penrose and Tegmark. On their view, reality is in essence *mathematical*. Mathematics, after all, is the science of structure; it neither knows of, nor cares about, stuff. Worlds that are structurally the same but made out of different stuff are identical in the eyes of the mathematician. Such worlds are called "isomorphic," from the Greek words *isos* (same) and *morphe* (form). If the universe is structure all the way down, then it can be exhaustively characterized by mathematics. And if mathematical structures have an objective existence, then the universe must *be* one of these structures. That, at least, seems to be Tegmark's meaning when he says that "all mathematical structures exist physically." If there is no ultimate stuff to reality, then mathematical structure is tantamount to physical existence. Who needs flesh when bones are enough?

A somewhat different take on a stuff-less reality is to see it as consisting not of mathematics but of *information*. This view is summed up in a slogan coined by the late physicist John Archibald Wheeler: "it from bit." (Wheeler—who collaborated with Albert Einstein and taught Richard Feynman—had a gift for such coinages; he also gets credit for "black hole," "wormhole," and "quantum foam.")

The it-from-bit story goes as follows. At bottom, science tells us only about *differences*: how differences in the distribution of mass/energy are associated with differences in the shape of spacetime, for example, or how differences in the charge of a particle are associated with differences in the forces it feels and exerts. States of the universe can thus be seen as pure information states. As the British astrophysicist Sir Arthur Eddington once put it, "Our knowledge of the nature of the objects treated in physics consists solely of readings of pointers on instrument dials." The "medium" in which these information states are realized, whatever it might be, plays no role at all in the explanation of physical phenomena. Therefore, it can be dispensed with altogether—shaved away by Occam's razor. The world is nothing but a flux of pure differences, without any underlying substance. Information ("bit") suffices for existence ("it").

Some it-from-bit proponents stretch this logic still further. They look on the universe as a giant computer simulation. Among those who have taken this view are Ed Fredkin and Stephen Wolfram, both of whom hypothesize that the universe is a "cellular automaton" that uses a simple program to generate complex physical outcomes. Perhaps the most radical cosmos-as-computer advocate is the American physicist Frank Tipler. The striking thing about Tipler's vision is that it involves no actual computer: his cosmos is all software, no hardware. A computer simulation, after all, is just the running of a program; and a program, in essence, is a rule that transforms an input string of numbers into an output string of numbers. So any computer simulation—say, the simulation of the physical universe—corresponds to sequences of strings of numbers: a pure mathematical entity. And if mathematical entities have an eternal Platonic existence, then, on Tipler's view, the existence of the world has been fully explained: "at the most basic ontological level," he declares, "the physical universe is a concept."

And what of the simulated beings who are somehow a part of that "concept"—beings like us? Would they realize that time was an illusion, that they were mere frozen bits of an eternal Platonic videotape? Not at all, according to Tipler. They would have no way of knowing that their reality consisted in being "a sequence of numbers." Yet, oddly enough, it is their simulated mental states that endow the overall mathematical concept of which they are a part with physical existence. For, as Tipler writes, "this is exactly what we mean by existence, namely, that thinking and feeling beings think and feel themselves to exist."

The picture of the universe as an abstract program—it from bit—strikes some thinkers as strangely beautiful. And it seems consistent with the way science represents nature, as a network of mathematical relations. But is that truly all there is? Is the world devoid of ultimate stuff? Is it indeed structure all the way down?

There is one aspect of reality that doesn't seem to have a place in this metaphysical picture: our own consciousness. Think of the way a pinch feels, a tangerine tastes, a cello sounds, or the rosy dawn sky looks. Such qualitative experiences—philosophers call them "qualia" (the plural of the Latin singular *quale*)—have an inner nature that goes beyond their role in the causal web. So, at least, philosophers like Thomas Nagel have argued. "The subjective features of conscious mental processes—as opposed to their purely physical causes and effects—cannot be captured by the purified form of thought suitable for dealing with the physical world that underlies appearances," Nagel writes.

One way of making this point vivid is due to the Australian philosopher Frank Jackson. Imagine, says Jackson, a scientist named Mary who knows everything there is to know about color: the neurobiological processes by which we perceive it, the physics of light, the composition of the spectrum, and so forth. But imagine further that Mary has lived her entire life in a black-and-white environment, that she has never actually *seen* a color herself. Despite her complete scientific understanding of color, there is something of which Mary is ignorant: what colors look like. She does not know what it is like to experience the color red. It follows that there is something to this experience—something subjective and qualitative—that is not captured by the objective, quantitative facts of science.

Nor, it would seem, can this subjective aspect of reality be captured by a computer simulation. Consider the theory called "functionalism," which holds that states of the mind are essentially computational states. What defines a mental state, according to functionalism, is not its intrinsic nature, but rather its place in a computer flowchart: the way it is causally related to perceptual inputs, to other mental states, and to behavioral outputs. Pain, for example, is defined as a state that is caused by tissue damage and that, in turn, causes withdrawal behavior and certain vocalizations, like "ouch." Such a flowchart of causal connections can be implemented in a software program, which, if run on a computer, would simulate *being in pain*.

But would this simulation duplicate what seems most real to us about pain: the horrible way it *feels*? To the philosopher John Searle, the very idea seems "frankly, quite crazy." "Why on earth," he asks, "would anyone in his right mind suppose a computer simulation of mental processes actually had mental processes?" Suppose, Searle says, the program simulating the experience of pain were to be run on a computer consisting of old beer cans tied together with string and powered by windmills. Can we really believe, Searle asks, that such a system would feel pain?

The philosopher Ned Block has come up with another thought experiment along the same lines. He invites us to imagine what would happen if the population of China were to simulate the brain's program. Suppose we got each Chinese person to mimic the activity of a particular brain cell. (There are only around a hundredth as many Chinese as there are cells in the human brain, but no matter.) Synaptic connections among the different cells could be simulated by cell-phone links among the Chinese. Would the nation of China, if it were to mimic the brain's software in this way, then have conscious states over and above those of its individuals? Could it experience, say, the taste of peppermint?

The conclusion the philosophers who come up with these thought experiments want us to draw is that there is more to consciousness than the mere processing of information. If this is true, then science, insofar as it describes the world as a play of information states, would seem to leave out a part of reality: the subjective, irreducibly qualitative part.

One could, of course, simply deny that reality *has* such a subjective part. And there are philosophers who do deny it—Daniel Dennett, for one. Dennett refuses to concede that consciousness contains any intrinsically qualitative elements. As far as he is concerned, "qualia" are a philosophical myth. If something cannot be described in purely quantitative and relational terms, it is simply not a part of reality. "Postulating special inner qualities that are not only private and intrinsically valuable, but also unconfirmable and uninvestigatable is just obscurantism," he declares.

Such denialism leaves philosophers like Searle and Nagel incredulous. It seems willfully blind to the very essence of what it means to be conscious. Nagel has written, "The world just *isn't* the world as it appears to one highly abstracted point of view"—that is, to the scientific point of view.

The inner nature of consciousness yields one reason for thinking that there is more to the world than pure structure. But apart from the issue of consciousness, there are more general grounds for suspecting that cosmic structuralism is inadequate as a picture of reality. Structure by itself just doesn't seem *enough* for genuine being. As the British idealist philosopher T. L. S. Sprigge put it, "What has structure must have something more to it than structure." Perhaps Aristotle was right: you need stuff too. Stuff is what gives existence to structure, what *realizes* it.

But if that is true, how can we come to have knowledge of the ultimate stuff of reality? Science, as we have seen, reveals only how the stuff is structured. It does not tell us how the quantitative differences it describes are grounded in differences in any underlying qualitative stuff. Our scientific knowledge of reality is thus, in Sprigge's words, "rather like the kind of knowledge of a piece of music which someone born deaf might have from a musical education based entirely on the study of musical scores."

Yet there is one part of reality that we do know without the mediation of science: our own consciousness. We experience the intrinsic qualities of our conscious states directly, from the inside. We have what philosophers call "privileged access" to them. There is nothing whose existence we are more certain of.

Now, this raises an interesting possibility. Maybe the part of reality we know indirectly through science, the physical part, has the same inner nature as the part we know directly through introspection, the conscious part. In other words, maybe all of reality—subjective and objective—is made out of the same basic stuff. That is a pleasingly simple hypothesis. But isn't it a bit crazy? Well, it didn't strike Bertrand Russell that way. In fact, it was essentially the conclusion Russell reached in *The Analysis of Matter*. Nor did it strike the great physicist Sir Arthur Eddington as crazy. In *The Nature of the Physical World* (1928), Eddington ringingly declared that "the stuff of the world is mind-stuff." (The term "mind-stuff," by the way, was coined by William James in the first volume of his 1890 work, *Principles of Psychology*.)

Crazy or not, the idea that the fundamental stuff of reality is mind-stuff has one very odd implication. If it is true, then consciousness must pervade all of physical nature. Subjective experience would not be confined to the brains of beings like us; it would be present in every bit of matter: in big things like galaxies and black holes, in little things like quarks and neutrinos, and in medium-sized things like flowers and rocks.

The doctrine that consciousness pervades reality is called "panpsychism." It seems to harken back to primitive superstitions like animism—the belief that trees and brooks harbor spirits. Yet it has attracted quite a bit of interest among contemporary philosophers. A few decades ago, Thomas Nagel showed that panpsychism, for all its apparent daftness, is an inescapable consequence of some quite reasonable premises. Our brains consist of material particles. These particles, in certain arrangements, produce subjective thoughts and feelings. Physical properties alone cannot account for subjectivity. (How could the ineffable experience of tasting a strawberry ever arise from the equations of physics?) Now, the properties of a complex system like the brain don't just pop into existence from nowhere; they must derive from the properties of that system's ultimate constituents. Those ultimate constituents must therefore have subjective features themselves—features that, in the right combinations, add up to our inner thoughts and feelings. But the electrons, protons, and neutrons making up our brains are no different from those making up the rest of the world. So the entire universe must consist of little bits of consciousness.

Another contemporary thinker who takes panpsychism seriously is the Australian philosopher David Chalmers. What attracts Chalmers to panpsychism is that it promises to solve two metaphysical problems for the price of one: the problem of stuff and the problem of consciousness. Not only does panpsychism furnish the basic stuff—mind-stuff—that might flesh out the purely structural world described by physics. It also explains why that otherwise gray physical world is bursting with Technicolor consciousness. Consciousness didn't mysteriously "emerge" in the universe when certain particles of matter chanced to come into the right arrangement; rather, it's been around from the very beginning, because those particles themselves are bits of consciousness. A single ontology thus underlies the subjective-information states in our minds and the objective-information states of the physical world—whence Chalmers's slogan: "Experience is information from the inside; physics is information from the outside."

If this metaphysical deal seems too good to be true, I should point out that panpsychism comes with problems of its own. Foremost among them is what might be called the Combination Problem: how can many little bits of mind-stuff combine to form a bigger mind? Your brain, for instance, is made up of lots of elementary particles. According to the panpsychist, each of these elementary particles is a tiny center of proto-consciousness, with its own (presumably very simple) mental states. Just what is it that makes all these micro-minds cohere into the macro-mind that is your own?

The Combination Problem proved a stumbling block for William James, who was otherwise friendly to panpsychism. "How can many conscious-nesses be at the same time one consciousness?" James asked in bewilder-ment. He made the point vivid with an example. "Take a sentence of a dozen words, and take twelve men and tell to each one word. Then stand the men in a row or jam them in a bunch, and let each think of his word as intently as he will; nowhere will there be a consciousness of the whole sentence. . . . The private minds do not agglomerate into a higher mind."

James's point is echoed by many contemporary opponents of panpsy-chism. What sense does it make, they say, to conjecture that things like electrons and protons are inwardly mental if you have no clue as to how their micro-mentality gets unified into full-blown human consciousness?

But there are a few intrepid thinkers who claim they do have a clue. And it is supplied, perhaps surprisingly, by quantum theory. One of the striking novelties of quantum theory is the notion of *entanglement*. When two distinct particles enter into a state of quantum entanglement, they lose their individual identities and act as a unified system. Any change to one of them will immediately be felt by the other, even if they are light-years apart. There is nothing analogous to this in classical physics. When quantum entanglement occurs, the whole becomes more than the sum of its parts. This is so radically at odds with our everyday way of viewing the world that Einstein himself pronounced it "spooky."

Now, even though quantum theory is customarily applied to a physical ontology, one consisting of particles and fields, there is no obvious rea-son why it can't also be applied to an ontology consisting of mind-stuff. Indeed, such a "quantum psychology" could hold the key to understanding the *unity* of consciousness—considered by Descartes and Kant to be a dis-tinctive mark of the mental. If physical entities can lose their individual identity and merge into a single whole, then it is at least conceivable that proto-mental entities could do likewise and—as William James put it—"agglomerate into a higher mind." Thus does quantum entanglement offer at least a hint of a solution to the Combination Problem.

Roger Penrose himself has invoked such quantum principles to explain how the physical activities in our brains generate consciousness. In *Shadows of the Mind*, he wrote that "the unity of a single mind can arise . . . only if there is some form of quantum coherence extending across an appre-ciable part of the brain." And he has since gone further, endorsing the panpsychist notion that the atomic constituents of the brain, along with the rest of the physical universe, are structured out of mind-stuff. "I think that something of this nature is indeed necessary," Penrose announced in a public lecture when the issue came up.

Panpsychism is not for everyone. John Searle, for one, dismisses it with-out argument as simply "absurd." But it has one undeniable virtue: that of ontological parsimony. It says that the cosmos is ultimately made of a single kind of stuff. It is thus a *monistic* view of reality. And if you are try-ing to solve the mystery of existence, monism is a convenient metaphysical position, since it obliges you to explain how only one substance came into being. The dualist has a seemingly harder job: he has to explain both why matter exists *and* why mind exists.

So does reality ultimately consist of mind-stuff? Is it no more (or no less) than an enormous, infinitely convoluted thought, or even dream? Seeking additional authority for this rather wild-sounding conclusion, I turned to what had hitherto proved an unimpeachable source: *The Devil's Dictionary*. There I found the following apt definition:

Reality, *n*. The dream of a mad philosopher.

11

"THE ETHICAL REQUIREDNESS OF THERE BEING SOMETHING"

"**W**ell, I have my pet answer, and I was very proud of it. But then, to my horror and disgust, I found that Plato had got the same answer about twenty-five hundred years ago!"

The man with the answer—one he believed to be utterly original when he first hit on it as a teenager—was a mild-mannered and soft-spoken speculative cosmologist by the name of John Leslie.

The community of speculative cosmologists is geographically scattered but not large. It consists of a hundred or so philosophically inclined scientists and scientifically adept philosophers—figures like Baron Rees of Ludlow, Britain's current Astronomer Royal; Andrei Linde, the Stanford physicist who created the theory of chaotic inflation; Jack Smart, the dean of Australian realist philosophy; and the Reverend Sir John Polkinghorne, a Cambridge particle physicist turned Anglican priest. In this far-flung and variegated community, John Leslie commands considerable respect—for both the boldness of his cosmic conjectures and the ingenuity with which he defends them. A native Englishman, Leslie took his graduate degree at Oxford in the early 1960s. He then moved to Canada, where he taught philosophy at the University of Guelph for three decades and was ultimately elected a fellow of the Royal Society. Over his career, he has produced a steady output of books and articles that blend technical rigor with conjectural fancy. His 1989 book, *Universes*, meticulously teased out the implications of the cosmic "fine-tuning" hypothesis for the existence of a multiverse. His 1996 book, *The End of the World*, showed how purely probabilistic reasoning pointed to a "doomsday" scenario in which humanity would be imminently extinguished. His 2007 book, *Immortality Defended*, drew on notions from contemporary physics—notably Einsteinian relativity and quantum entanglement—to argue that, biological death notwithstanding, there is a very real sense in which each of us will exist eternally. As a recreational sideline, Leslie invented a new board game called "Hostage Chess." A blend of Western chess and the Japanese game of Shogi, Leslie's Hostage Chess has been called by one grandmaster "the most interesting, exciting variant that can be played with a standard chess set."

For all that, the achievement for which Leslie says he wishes to be remembered is his proposed solution to the mystery of why there is Something rather than Nothing—even if, as he concedes, Plato beat him to it. (Well, didn't Alfred North Whitehead say that all philosophy was a footnote to Plato?) He calls his solution "extreme axiarchism," since it holds that reality is ruled by abstract value—*axia* being the Greek word for "value" and *archein* for "to rule."

"You're the world's foremost authority on why there is Something rather than Nothing," I said to Leslie at the outset of our conversation. He was sitting in the living room of his house on the west coast of Canada, comfortably attired in a wool crewneck against the late-fall chill, while I hovered about in the noosphere.

"I doubt that there's any sort of *authority* on why the world exists," he replied, waving a hand and blinking behind his spectacles. "I'm an authority on the range of guesses which have been given. But I do have my own ideas, which, as I said, go back to Plato. Plato thought that there was a necessarily existing realm of possibilities, and I believe he was right about that."

Existing possibilities?

"Well," Leslie said, "even if nothing at all existed, there would still be all sorts of logical possibilities. For instance, it would be true that apples—unlike married bachelors—were logically possible, even though they did not actually exist. It would also be true that *if* two sets of two apples were

to exist, *then* there would exist four apples. Even if there had been nothing at all, such conditional truths, truths of an *if*-y *then*-y sort, would still
have held."

Fine, I said, but how do you get from such possibilities—from "*if*-y *then*-y
truths," as he called them—to actual existence?

"Well," Leslie resumed, "Plato looked among these truths and recognized that some of them were more than just *if*-y *then*-y. Suppose you had an
empty universe—nothing at all. It would be a fact that this empty universe
was a lot better than a universe full of people who were in immense misery.
And this would mean that there was an ethical *need* for the emptiness to
continue rather than being replaced by a universe of infinite suffering. But
there might also be another ethical need in the opposite direction—a need
for this emptiness to be replaced with a *good* universe, one full of happiness
and beauty. And Plato thought that the ethical requirement that a good
universe exist was itself enough to *create* the universe."

Leslie called my attention to Plato's *Republic*, in which we are told that
the Form of the Good is "what bestows existence upon things." Leslie's
own answer to the puzzle of existence, he said, was essentially an updating
of that Platonic claim.

"So," I said, trying to sound less incredulous than I felt, "you're actually suggesting that the universe somehow exploded into being out of an
abstract need for goodness?"

Leslie was coolly unflappable. "Provided you accept the view that this
world is, on balance, a good world, the idea that it was created by the *need*
for the existence of a good world can at least get off the ground," he said.
"This has persuaded a lot of people over the ages since Plato. For those
who believe in God, it has even provided an explanation for God's *own*
existence: he exists because of the ethical need for a *perfect being*. The
idea that goodness can be responsible for existence has had quite a long
history—which, as I've said, was a great disappointment for me to discover,
because I'd have liked it to have been all my own."

Something in Leslie's soft, precise diction, which always betrayed just
a hint of mirth, made me suspect there might be an undercurrent of irony
in his Platonic creation story. And if he was seriously claiming that the

universe sprang into being in answer to an ethical need for goodness, then could he explain why it has turned out to be such a disappointment, ethically and aesthetically speaking—howlingly mediocre when not downright evil?

It was then that I learned that reality according to Leslie far outstrips reality as the rest of us know it.

To begin with, if existence arose out of a need for goodness, then it must be essentially *mental*. In other words, existence must ultimately consist of mind, of consciousness. The reason, according to Leslie, is simple. For something to be valuable *in itself*, as opposed to being valuable as a means to an end, that thing must have unity. It must be more than just an assemblage of separately existing parts. Granted, you can make something that is instrumentally valuable by putting together valueless parts—a TV set, for example. A TV set has instrumental value because it can produce enjoyment in someone watching it. But the experience of enjoyment is a state of consciousness. It has a unity that goes beyond any merely mechanical organization of parts. And that is why such a conscious experience can be *intrinsically* valuable. It was G. E. Moore—the founder, along with Bertrand Russell, of modern analytic philosophy—who first laid stress on the crucial role of what he called "organic unity" in the existence of intrinsic value. And genuine organic unity—as opposed to mere structural unity, the unity of an automobile engine or a heap of sand—is realized only in consciousness. (As William James observed, "However complex the object may be, the thought of it is one undivided state of consciousness.") So if the world was indeed ushered into being by a need for goodness, then it must be fundamentally *made* out of consciousness.

That much, at least, I had gleaned from Leslie's earlier writings, like his 1979 book, *Value and Existence*. What I was not prepared for was the great enlargement that his cosmic scheme had undergone in the intervening years.

"In my grand vision," he told me, "what the cosmos consists of is an *infinite* number of *infinite* minds, each of which knows *absolutely everything* which is worth knowing. And one of the things which is worth knowing is the structure of a universe such as ours."

So the physical universe itself, with its hundreds of billions of galaxies, is just the contemplative product of one of those infinite minds. That was what Leslie was telling me. And the same goes for the inhabitants of the universe—us—and their conscious states. So my question remained, If an infinite mind was thinking the whole show up, why all the evil and suffering and disaster and sheer ugliness? Why do we inhabit such a darkling plain?

"But our universe is just *one* of the structures that an infinite mind would contemplate," he said. "It would also know the structure of infinitely many *other* universes. And it would be very unlikely for ours to be the *best* of all of them. The best situation is the *total* situation, with all of these vastly many universes coexisting as contemplative patterns in an infinite mind. And the perfectly beautiful universe that you'd prefer? Well, maybe it's one of those contemplative patterns. But there's also *our* universe as well. I suspect that, of all the infinitely many worlds that are being thought of by an infinite mind, we're pretty far down the list in terms of overall goodness. Still, I think you'd have to go quite far below us to have a world which was not worth having *at all*."

Here Leslie chuckled audibly. Then, recovering his graver demeanor, he invited me to consider the Louvre museum as an analogy. Just as an infinite mind contains many universes, the Louvre contains many artworks. One of these artworks—say, the *Mona Lisa*—is the best. But if the Louvre contained nothing but perfect replicas of the *Mona Lisa*, it would be a less interesting museum than it actually is, with its vast number of inferior artworks adding to the variety. The best museum on the whole is one that contains, in addition to the very best works of art, all lesser works, as long as those lesser works have some redeeming aesthetic value—as long, that is, as they are not positively bad. Similarly, the best infinite mind is one that contemplates all cosmic patterns whose net value is positive, ranging from the very best possible world on down to worlds of indifferent quality, where the good barely outweighs the evil. Such a variety of worlds, each of which is, on the whole, better by some positive margin than sheer nothingness, is the most valuable reality overall—the one that might leap into existence out of a Platonic requirement for goodness.

Leslie had answered one obvious objection to his cosmic scheme: the problem of evil. Our own world is decidedly not the *Mona Lisa*. It is blemished by cruelty, suffering, arbitrariness, and waste. Yet, even with all its ethical and aesthetic defects, it manages to contribute a little net value to reality as a whole—just the way a mediocre painting by a second-rate artist might contribute a little net value to the collection in the Louvre. Our world is thus worthy to be part of that larger reality: worthy, that is, of contemplation by an infinite mind.

But there remained a still graver objection to Leslie's axiarchic theory. Why should an infinite mind—or anything else, for that matter—be summoned into existence by a sheer need for goodness in the first place? Why, in other words, should "ought to exist" imply "does exist"? Such a principle certainly doesn't seem to operate in the real world. If a poor child is starving to death, it would be good if a bowl of rice were to come into existence to save that child's life. Yet we never see a bowl of rice materialize for the child out of nothingness. So why should we expect an entire cosmos to do the same?

When I put this objection to Leslie, he emitted a long sigh.

"People like me," he said, "people who accept the Platonic view that the universe exists because it ought to exist, we aren't saying that absolutely *all* ethical requirements are satisfied. We recognize that there are conflicts. If you're going to have an orderly world that runs according to laws of nature—which is a very elegant and interesting way for a world to be—you can't have bowls of rice suddenly appearing miraculously. Moreover, the fact that the child doesn't have a bowl of rice may very well be the result of a misuse of human freedom, and you can't have the goodness of a world where agents are free to make decisions unless you also have the possibility that those agents will make *bad* decisions."

I understood that the requirements of goodness could conflict, that some could be overruled by others. But why should goodness have any tendency to fulfill itself at all? Why should it be different from, say, redness? Redness clearly doesn't have a tendency to fulfill itself. If it did, everything would be red.

"Richard Dawkins once made the same point. He asked me, 'How could so piffling a concept as goodness explain the world's existence? You might as well appeal to Chanel Number Fiveness.' Well, I don't think of goodness as just another quality that is slapped on things like perfume or a coat of paint. Goodness is *required existence,* in a nontrivial sense. Anyone who doesn't grasp that hasn't reached square one in understanding what ethics is all about."

Imagine some good possibility—like that of a beautiful and harmonious cosmos just spilling over with happiness. If that possibility were made real, it would have an existence that was ethically needful. This was essentially Plato's idea: that a thing could exist *because* its existence was required by goodness. The connection between goodness and required existence isn't a logical one. Yet it *is* a necessary connection—that, at least, is what Platonically inclined thinkers like Leslie believe. We may simply lack the conceptual resources to appreciate why this is so. We tend to think that value can bring something into existence only with the aid of some *mechanism*—as Leslie put it, "some combination, perhaps, of pistons pushing, electromagnetic fields tugging, or persons exerting willpower." But such a mechanism could never explain the existence of a world. It could never explain why there is Something rather than Nothing, because it would be part of the Something to be explained. Given the limitations of our understanding, we have to content ourselves with the bare insight that an ethical need and a creative force both point in the same direction: toward Being. The Platonic idea that there is a necessary connection between the two is not an inescapable truth of logic. But neither is it a conceptual absurdity. So, at any rate, Leslie was maintaining.

Perhaps, I suggested to him, it might help to think of the matter the other way around. Even if an abstract need for goodness did not in itself provide a very compelling reason for a cosmos to exist, it at least provided *some* reason. And in the absence of a countervailing reason—a reason that would oppose the existence of the world—goodness alone might be enough to secure the victory of Being over Nothingness. From a physical point of view, after all, the universe doesn't seem to cost anything: its total energy, when the negative gravitational energy is balanced against the positive energy locked up in matter, is zero.

Leslie welcomed this reasoning. "In the absence of a nihilistic force fighting the existence of things," he said, "*any* valid reason for their existence would tend to bring about their realization. You might dream up a sort of demon that was opposing the existence of things. But then, I ask, where did that demon come from?"

What about Heidegger, though? Didn't he believe in an abstract annihilating force? The Nothing that "noths"?

"Maybe he did, but I don't," Leslie replied. "If you actually read Heidegger, he's very obscure on the question of explaining existence. But he's been interpreted by the theologian Hans Küng as holding that the word 'God' is just a label for a creative ethical principle that's producing the world. So Heidegger may well be in the Plato-Leslie camp!"

Leslie himself, for all his theologically flavored talk of "divine minds," had little sympathy for the traditional concept of God. "If my view is true," he said, "what you are stuck with is an infinite number of infinite minds, each of which knows absolutely everything worth knowing. You can call each one of them 'God' if you want, or you could say that God was the entire infinite collection. Or you could even say that God was simply the abstract principle behind them all."

I recalled an observation that the orthodox Christian philosopher Richard Swinburne had made when I talked to him in Oxford. God can't be an abstract principle, Swinburne had insisted, because an abstract principle cannot suffer. And, when we suffer in a good cause, our creator has an obligation to suffer along with us, the way a parent has an obligation to suffer along with a child. The world would be a less good place if it weren't created by a God who shared our suffering—so Swinburne had claimed. And an abstract principle of goodness can't do that.

"Hmm," said Leslie very slowly. "That sounds like an argument for the existence of a Supreme Masochist. I find it hard to swallow the notion that the world is *improved* by extra suffering. And that goes for a lot of Christian doctrine. Jones commits a crime, so you expiate the evil by nailing Smith to a cross and it's all better."

Perhaps Leslie was more a pantheist then, in the style of Spinoza. Spinoza's God was not a personal agent, like the traditional deity of

Judeo-Christianity. Rather, Spinoza equated God to an infinite and self-subsistent substance that encompassed all of nature.

"A lot of people thought Spinoza wasn't talking about God at all," Leslie said. "They called him an atheist. And if you want to call me an atheist, that's fine by me. Words like 'theism' and 'atheism' and 'God,' they've moved around so much that they're practically meaningless. Who really cares? I do consider myself a Spinozist, however, for two reasons. First, I think Spinoza was right that we're all tiny regions in an infinite mind. And I agree with him that the material world, the world described by science, is a pattern of divine thought. But I also think that Spinoza himself was really a Platonist. That's not the standard view, of course. In his *Ethics*, Spinoza argues that the world exists as a matter of logical necessity. But the *Ethics* was not Spinoza's best book. His best book was an earlier one, *A Short Treatise on God, Man, and His Well-Being*. And there Spinoza pretty clearly runs the view that it is *value* that is creating everything—that the world exists because it's *good* that it should. When he got to *Ethics*, he wanted to prove everything in geometrical fashion, so he gave what looks like a logical proof, and not a very convincing one, that there must be an infinite substance. Consistency is the virtue of small minds, and Spinoza had a great mind—he was inconsistent all over the place."

Whether Platonic or Spinozistic, Leslie's view of reality had a certain beauty about it, I thought: the beauty of an ontological pipe dream. Yet, for all the rigor of his arguments—and he was never at a loss for an argument to rebut any objection—could his axiarchism (value rules!) really be taken seriously as the ultimate explanation for all existence?

As I was to discover, many thinkers have taken it quite seriously. Among them was the late Oxford philosopher (and staunch atheist) John Mackie. In his powerful book-length case against the existence of God, *The Miracle of Theism*, Mackie devoted an entire chapter, titled "Replacements for God," to Leslie's axiarchism. "The notion that the mere ethical *need* for something could *on its own* call that item into existence, without the operation of any person or mind that was aware of this need and acted so as to fulfil it, is, no doubt, initially strange and paradoxical," Mackie wrote. "Yet in it lies also the great strength of extreme axiarchism." Leslie's theory, he went on to say, "offers the only possible answer to the question which underlies all forms of the cosmological argument, the question 'Why is there anything at all?' or 'Why should there be any world rather than none?'"

Obviously, Mackie observed, no explanation in terms of a "first cause" could answer the ultimate question of existence, for such an explanation would merely raise the further question of why that first cause—whether it be God, an unstable chunk of false vacuum, or some other still more exotic entity—itself existed. But Leslie's explanation for the existence of the world did not have this defect, Mackie observed. The objective need for goodness that he posits is not a cause. It is rather a *fact*, a necessary fact, one that does not call for any further explanation. Goodness is not an agent or a mechanism that creates something out of nothing. It is a *reason* for there being a world rather than nothingness. In the end, though, Mackie remained skeptical of Leslie's axiarchism. He was not convinced that "something's being valuable can in itself tend to bring that thing into existence."

And neither was I. Metaphysics is all very nice, I said to Leslie, but what hard evidence did he have for his extraordinarily speculative claim about the existence of the world?

He reacted with barely disguised exasperation: "I'm always a little astonished when people say, 'Look, there's no evidence for your view.' Well, I say, there's one rather striking piece of evidence: *the fact that there is a world rather than just a blank*. Why do they discount this? The sheer existence of something rather than nothing simply cries out for explanation. And where are the competitors to my Platonic theory?"

Well, he had a point there. So far, at least, none of the other solutions I had heard proposed—those based on quantum cosmology, or on mathematical necessity, or on God—had held up. At this point, Platonic goodness appeared to be the only cosmic suspect out there.

Still, there seemed to be something circular about Leslie's use of evidence. The world was brought into existence by goodness. And how do we know that goodness can bring a world into existence? Because the world exists! If axiarchism was to amount to more than an empty tautology, Leslie was going to have to produce some additional evidence in its favor—something beyond the sheer existence of the world.

And so he did.

"A further bit of evidence is that the world is full of orderly patterns," he said. "Why does the universe obey causal laws? And why laws of such simplicity, rather than vastly more complex ones? In the last century, philosophers of science have doubted whether the causal orderliness of the universe could ever be explained. But it does seem to *need* an explanation. After all, order is improbable, not to be expected. There are so many more ways for a world to be a complete mess than to be nice and orderly. So why do elementary particles perform their mathematically elegant pirouettes? For a Platonist like me, such regularities are accounted for in the same manner that the presence of something rather than nothing is accounted for—by their ethical requiredness."

"Causal orderliness" seemed to be more of an aesthetic value than an ethical one, I noted.

"I've never been able to see the difference between the two," Leslie said. "All value is about what *ought to exist*. By the way, there's a third bit of evidence for my Platonic theory: the fact that the fundamental constants of nature are fine-tuned for intelligent life."

But, I objected, couldn't this appearance of cosmic fine-tuning be explained by science? Suppose, as physicists like Steven Weinberg believe, our universe is but one region of a multiverse. Suppose further that the constants of nature take different values in different regions of this multiverse. Then, by the anthropic principle, isn't it to be expected that we should find ourselves in a region where those constants are favorable to the evolution of beings like us? No need for Plato when you have a multiverse!

"There are a couple of ways I can react to that," Leslie said. "The fact that the multiverse hypothesis is an alternative to the axiarchic hypothesis doesn't mean that *both* of them can't be strengthened by evidence of fine-tuning. Let me tell you a little parable, the parable of the vanished treasure. You're on a desert island, and you've buried a treasure chest there. The only other people on the island are Smith and Jones. One day you go to the place where you buried the treasure chest, and you try to dig it up. And it's not there! Now, the fact that it's not there increases the probability that Jones is a thief, but it also increases the probability of the competing hypothesis that Smith is a thief. In the same way, the discovery of cosmic fine-tuning strengthens the probability that the multiverse hypothesis is correct, but it also strengthens the probability that my axiarchic hypothesis is correct."

He went on to make a far more subtle point—a point that, as far as I could tell, was entirely original: the multiverse hypothesis doesn't really solve the mystery of fine-tuning at all.

"Notice," Leslie said, "that for life to evolve in the universe, each of the cosmic constants needs to be fine-tuned in a particular way *for many differ-ent reasons at once*. The strength of the electromagnetic force, for example, has to be in a particular narrow range, *first*, so matter would be distin-guished from radiation and you have something to make living beings out of; *second*, so that all quarks wouldn't turn into leptons, meaning there never would have been any atoms; *third*, so that protons wouldn't decay so quickly that there'd soon be no atoms remaining, let alone organisms to survive the radiation produced by the decay; *fourth*, for protons not to repel one another so strongly that there'd be no such thing as chemistry, and hence no chemically-based beings like us."

He continued with a *fifth*, a *sixth*, a *seventh*, and an *eighth* reason, each of greater technical complexity.

"Now," Leslie said, having concluded the litany, "how is it that one and the same twiddling of the cosmic knob for the strength of the electromag-netic force should satisfy so many requirements? This doesn't seem to be a problem that can be solved by the multiverse model. The multiverse model only says that the strength of the electromagnetic force varies by chance from universe to universe. But for even a *single* life-permitting strength to be possible, the fundamental laws of physics themselves have to be *just so*. In other words, those laws—which are, by the way, supposed to be the same all across the multiverse—must have *the potential for intelligent life built into them*. Which is precisely why they would be the sort of laws that an infinite mind might find it interesting to contemplate."

It was an awfully tidy package, Leslie's axiarchism. Whatever you thought of its mind-bending assumptions—the Platonic reality of good-ness, the creative efficacy of value—you had to admire its completeness and coherence as a speculative construction. And I did admire it. But I wasn't quite *moved* by it. It didn't quite speak to my existential depths. It didn't appease my hunger for ultimate explanation. In fact, I wondered how deeply Leslie himself was invested in it, emotionally speaking. Did he feel anything like a quasi-religious attachment to his theory?

"Um ... uh ... um ... ," he stammered, sounding almost pained. "I feel constantly *embarrassed* by the idea that I ought to be attracted to my system because, well, wouldn't it be lovely if it were true. That is just pie in the sky, and I very much dislike it. I don't have anything like *faith* in my Platonic creation story. I certainly haven't *proved* its truth. Almost nothing of philosophical interest strikes me as being provable. I'd say my confidence in it is just a little over 50 percent. A lot of the time, I feel that the universe just happens to exist and that's it."

Was the possibility that the world might exist for no reason whatever disturbing to him?

"Yes," he replied, "it is—on an intellectual level, at least."

Still, I added, he must find it gratifying that a significant minority of other philosophers have come around to his view.

"Or to other views that are equally crazy," he said.

WAS LESLIE'S AXIARCHISM the long-sought resolution to the mystery of existence? Had the answer to the question *Why is there something rather than nothing?* been available virtually from the beginning of Western thought, in the form of Plato's vision of the Good? If so, why did so many subsequent thinkers—Leibniz, William James, Wittgenstein, Sartre, and Stephen Hawking, to name a few—fail to see it? Were they all prisoners in Plato's cave?

To take axiarchism seriously, you have to believe three things.

First, you have to believe that goodness is an objective value—that there are facts about what is good and evil, and that these facts are timelessly and necessarily true, independently of human concerns, and that they would be true even in the absence of all existent things.

Second, you have to believe that the ethical needs that arise from such facts about goodness can be creatively effective—that they can bring things into existence and maintain those things in existence without the aid of any intermediary agent or force or mechanism.

Third, you have to believe that the actual world—the world that we ourselves are a part of, even if we can only see a very tiny region of it—is the sort of reality that abstract goodness would bring into being.

In other words, you have to believe that (1) *value is objective*, (2) *value is creative*, and (3) *the world is good*. If you buy into all three of these propositions, you've got your resolution to the mystery of existence.

The first of them is philosophically controversial, to say the least. The most radical of the value skeptics, who trace their lineage to David Hume, hold that there is no such thing as objective goodness. Our judgments of right and wrong, on the Humean view, are just a matter of our sentiments, which we project onto the world and imagine to be part of the fabric of reality. Such moral judgments have nothing to do with objective truth, or even with reason. As Hume himself famously put it, " 'Tis not contrary to reason to prefer the destruction of the whole world to the scratching of my finger."

That is surely taking skepticism about value too far. Yet even philosophers on the opposite side of the issue, those who staunchly affirm the objectivity of value, have doubts about whether ethical needs could float completely free of the interests and concerns of sentient beings like ourselves. As Thomas Nagel once asked, if all sentient life were destroyed, would it still be a good thing if the Frick Collection survived?

Leslie himself is what might be called an "objective subjectivist" when it comes to value. He is a *subjectivist* because he believes that value ultimately resides only in states of consciousness, not in anything outside the mind. Yet he is an *objective* subjectivist because he believes that happiness is objectively better than suffering, not better merely because we happen to prefer it.

Why is a world of happy sentient beings objectively better than nothingness? Well, you might say, if there were a world of happy sentient beings, its annihilation would be ethically bad. But suppose we start from the nothingness side. If there were nothing at all, would it be objectively better if a world of happy sentient beings were to pop into existence?

Perhaps it would. After all, the sum of happiness would go from zero to some positive number, which seems an objectively good thing. And it also seems objectively true that the sentient beings brought into existence were thereby benefitted (although it would be odd to say that if those sentient beings had *not* been brought into existence, they would thereby have been *harmed*).

But—moving on to the second point—even if there are objective truths about goodness, how could those truths *do* anything? How could they summon up a world out of sheer nothingness? Even if values are objective, they are not "out there" the way galaxies and black holes are. (If they were, they'd be useless for explaining why there is something rather than nothing, because they would be part of the something to be explained.) To say that values are objective is to say that we have objective *reasons* to do certain things. And reasons require agents to act on them if they are to have any impact on reality. Reasons without agents are impotent. To believe otherwise is to flirt with the scientifically discredited Aristotelian notion of "final cause" or "immanent teleology"—that it rains in the spring because that is good for the crops.

But perhaps this conclusion is too hasty. Can we make sense of a reason that might favor the existence of something even in the absence of any person who might act on the reason—a reason not to *do* but to *be*? What we're looking for, remember, is an explanation of why there is anything at all—a causal explanation. Now, what styles of causal explanations are there? Well, there is *event* causation, where one event (say, the decay of a certain scalar field) causes another (the Big Bang). And there is *agent* causation, where an agent (say, God) causes an event (the Big Bang). Evidently, neither event causation nor agent causation can explain why there is something rather than nothing, since each presupposes the existence of something. There is, however, a third style of causal explanation, *fact* causation, where the fact that p causally explains the fact that q. In most cases of fact causation we are familiar with, the causing fact p involves something that exists—as in, for instance, "Jones died because he swallowed poison." Yet it may be that, when q is the fact that *there is something rather than nothing*, the causing fact p needn't itself cite anything that exists—any agent or substance or event. The causing fact might just be an abstract *reason*. And, if there is no additional fact that opposes or undermines this abstract reason, then such a reason could make for an adequate causal explanation. That, indeed, would appear to be the only hope for a noncircular resolution to the mystery of existence.

However—and now we move on to the third part of the axiarchic case—is it really plausible that the explaining reason should be that this world is better than an ontological blank? Actually, the axiarchist is committed to a much stronger thesis. He must believe that the world is not merely *better* than nothing, but that it is *maximally* good, *infinitely* good, the nicest reality that money can buy.

Ever since Leibniz made the fatuous-sounding claim that we live in the "best of all possible worlds" (and was mercilessly mocked by Voltaire for doing so), apologists for the goodness of creation have tried to explain away the apparent evil that permeates it. Perhaps, they say, evil has no genuine reality but is merely a negation, the local absence of goodness, the way blindness is the absence of sight. (This is the so-called privative theory of evil.) Or perhaps evil is an inevitable by-product of the good of freedom, which cannot exist without the possibility that it will be abused. Or, again, perhaps a bit of evil makes reality better as an "organic whole"— the way the dissonance in a Mozart string quartet heightens its overall beauty, or the way death is necessary to the aesthetic power of tragedy. After all, a world that is good through and through is a bland world; it is the presence of evils to be overcome through noble struggle that gives it piquancy. And sometimes evil itself can come to seem positively glamorous and romantic. What would *Paradise Lost* be without the rebellious pride of Satan?

Leslie himself concedes the existence of evil. He admits that "many items in our universe are far from splendid"—ranging from headaches to mass murder to the destruction of entire galaxies through false-vacuum fiascoes. Yet he purports to render the problem of evil manageable by making our world a tiny part of a much greater reality—a reality consisting of an infinite number of infinite minds, each of them contemplating everything of value. As long as the world around us contributes at least a little net value to this infinite reality, its existence is sanctioned by the abstract need for goodness. It may not be perfect, but—with its causal orderliness, its congeniality to life, and its conduciveness to more happy states of consciousness than unhappy ones—it's good enough to merit inclusion in a maximally valuable reality.

So, at least, Leslie claimed. Yet I wondered whether he wasn't just projecting his own happy consciousness onto a harsh and uncaring cosmos. He struck me as a temperamentally sunny man, one whose capacity for skepticism and irony only enhanced the intellectual pleasure he took in the worldview he had so painstakingly elaborated. In fact, he struck me as a sort of latter-day Spinoza. Leslie's own metaphysical scheme, as he cheerfully admitted, was Spinozistic in flavor (even if, with its infinite number of pantheistic minds, it was "far richer" than the one that Spinoza described). Like Spinoza, Leslie sees all individual things as ripples on the sea of a unified divine reality. By all accounts, Spinoza was endowed with a deep intellectual reverence for this reality. His gentle integrity made him, according to Bertrand Russell, "the noblest and most lovable of the great philosophers." Spinoza understood human suffering—of which he experienced his share, being ostracized as an infidel by his fellow Jews, and as a dangerous atheist by Christians—to be a minor discord in a larger cosmic harmony. Leslie seemed to have the same gift. And, like Spinoza, he lived as a sort of exile—in Canada.

It is tempting to join the sunny Spinoza-Leslie consensus. There is something to be said for cosmic optimism—especially when it not only helps us avoid despair in the face of evil, but also promises to explain why the world exists. But there is also something to be said for the contrary point of view. Schopenhauer said it in the nineteenth century: reality is overwhelmingly a theater of suffering, and nonexistence is better than existence. So did Byron, in his lines, "Sorrow is knowledge: they who know the most / Must mourn the deepest o'er the fatal truth. . . ." More recently, Camus declared that the only genuine philosophical problem is suicide, and E. M. Cioran epigrammatized endlessly about the "curse" of existence. Even Bertrand Russell, despite his professed admiration for Spinoza's character, could not accept the Spinozist view that individual evils are neutralized by absorption into a larger whole. "Each act of cruelty," Russell insisted, "is eternally a part of the universe." Today, the most uncompromising opponent of cosmic optimism may be Woody Allen. In an interview he gave in 2010 (to a Catholic priest, curiously enough), Allen spoke of the "overwhelming bleakness" of the universe. "Human existence is a brutal experience to me," he said. "It's a brutal, meaningless experience—an agonizing, meaningless experience with some oases, delight, some charm and peace, but these are just small oases." There is no justice to it, Allen maintained, and no rationality either. Everyone does what one can to alleviate "the agony of the human condition." Some distort it with religion; some chase money or love. Allen himself makes films—and whines. ("I do get a certain amount of solace from whining.") Yet in the end "everyone goes to his grave in a meaningless way."

A confirmed axiarchist might respond that Woody Allen takes too parochial a view of reality. There is more to heaven and earth than is encompassed in the morbid imagination of a Manhattan neurotic. But it could be argued that it is John Leslie, in his still hearth among the barren crags of Canada's western coast, far from all centers of civilization, who is the cosmically parochial one. Leslie cites the causal orderliness of the universe and its fine-tuning for life as self-evidently good, as things that *ought to be*. But do they outweigh the sheer volume of agony inflicted on sentient beings, often by one another?

Maybe Leslie is right about one thing. Maybe the world really does owe its existence to some sort of abstract principle. But it seems unlikely that this principle should be intimately bound up with human concerns and judgments, the way goodness is. Leslie's "creative value" looks too much like the ghost of a Judeo-Christian deity, a deity that we made in our own image and likeness. Could there be some other Platonic possibility, perhaps stranger and more alien to us, that might be behind the existence of the world, that might explain why there is Something rather than Nothing? To find a fitting resolution to the mystery of existence, I'd have to broaden the search. And, as it turned out, I'd have to get comfortable with a new and unfamiliar notion: "the Selector."

Before taking leave of Leslie, though, I wanted to salute him for producing a play of ideas that was so consistently enlightening—and, not incidentally, entertaining.

"Of all the contemporary philosophers I've been reading," I told him, "you've got to be the wittiest."

"You're very kind," he said. Then he added, "But I'm not sure that's much of a compliment."

Interlude

An Hegelian in Paris

*P**ure Being makes the beginning . . .***

I read these words while sitting—yet again—at a table at the Café de Flore. This time I am on the terrace of the café, facing the busy Boulevard Saint-Germain and, across the street, the Brasserie Lipp, with its promise of *choucroute garnie*. It is one of those rare early-spring days when the delicate oyster-shell gray of the Parisian sky gives way to an access of brilliant sunshine and cobalt blue. Distracted by the lovely weather, I look up from the page for a moment, hoping that I might spot an acquaintance, or at least a recognizable face, among the parade of people passing to and fro along the broad sidewalk in front of me. *Pas de veine.* So I sip the last bit of the café *express* I ordered—my fourth since I've been here—and return to my book, which happens to be Hegel's *Science of Logic*.

That may seem an odd, not to say pretentious, choice of reading material for an idle afternoon in a fashionable (and overpriced) Left Bank café. But it isn't odd, really. I am, after all, in a place that Jean-Paul Sartre and Simone de Beauvoir made their daily headquarters some decades ago. It was here, in the winter of 1941–42, during the German occupation of Paris, that Sartre began composing his most imposing philosophical treatise, *Being and Nothingness*. That winter was a brutally cold one, but the proprietor of the café, Monsieur Boubal, was adept at procuring enough black-market coal to keep the interior at least minimally heated, and enough tobacco to supply the wants of its smoking patrons. Sartre and de Beauvoir would typically show up first thing in the morning and install themselves at the warmest table, next to the stove pipe. Sartre would ask for a cup of tea with milk, his sole order for the entire day. Then, still bundled up in his fake-fur coat of bright orange, and wearing his round horn-rim glasses, he would scribble away for hours at a stretch, barely looking up from the paper—except (as Beauvoir recalled in her memoirs) to retrieve from the floor and stuff into his briar pipe the occasional cigarette butt discarded by another customer.

And how did Sartre begin his epic inquiry into the relationship between *l'être et le néant*? With a description of this very café as "a fullness of being"—followed by a lengthy riff on the dialectic of being that Hegel set out in his *Logic*. So it is hardly incongruous that I should be striking a Hegelian pose here. As for pretentious . . . well, Café de Flore sets a very high bar for pretension.

My purpose, though, is a serious one. What I am struggling to do is to see the world in the most abstract way possible. That, it seems to me, is the best remaining hope for puzzling out why the world exists at all. All of the thinkers I had already spoken to fell short of complete ontological generality. They saw the world under some limited aspect. To Richard Swinburne, it was a manifestation of divine will. To Alex Vilenkin, it was a runaway fluctuation in a quantum vacuum. To Roger Penrose, it was the expression of a Platonic mathematical essence. To John Leslie, it was an outcropping of timeless value. Each of these ways of seeing the world purported to yield the answer to why it exists. But none of these answers struck me as satisfactory. They didn't penetrate to the root of the existential mystery—to what Aristotle, in his *Metaphysics*, called "being *qua* being." What does it mean *to be*? Is being a kind of property, one possessed by all existing things? Is it an activity, as the participial form of the word "be-ing" suggests? Clearly, one could not expect to understand *why* there is being without first having some grasp of *what* being really is.

And so, like Sartre before me, I find myself turning to Hegel. His doctrine of pure being has been one of the most influential in the whole history of philosophy—that much I knew. And it is in his *Logic* that he reputedly laid out this doctrine in its most comprehensible form.

"Pure Being makes the beginning," Hegel declares at the outset, "because it is on the one hand pure thought and on the other immediacy itself."

So far, so good, I think. You can't get really anywhere in your philosophizing without acknowledging that there *is* something.

But what can we say about this Pure Being? Well, at its very purest, Hegel observes, it is "simple and indeterminate." It has no specific qualities, such as number, size, or color.

That too also makes sense. Pure being is not like an apple, or a golf ball, or a dozen eggs.

Very quickly, though, Hegel's reasoning takes a peculiar turn. "This mere Being, as it is mere abstraction, is therefore the absolutely negative," he avers. In other words, since Pure Being is the absence of all qualities, it is equally the *negation* of all qualities.

And what follows from this? That Pure Being "is just Nothing."

Did I hear a rim shot?

Hegel is aware of the apparent absurdity of this conclusion. "No great expenditure of wit is needed to make fun of the maxim that Being and Nothing are the same," I read. Nevertheless, the two concepts, at this rarefied level of abstraction, are identically empty. Each thus harbors the other within itself. They are dialectical twins.

Yet, despite their conceptual twinhood, Being and Nothing remain mutually contradictory. They stand in opposition to each other. Therefore, Hegel observes, they have to be reconciled. They must be brought together into a unity, a unity that supersedes these two timeless categories without destroying their distinctness.

And what is it that heals the breach? Becoming!

Thus does the great Hegelian dialectic get under way. *Thesis*: Reality is Pure Being. *Antithesis*: Reality is Nothing. *Synthesis*: Reality is Becoming.

Pure Becoming would seem as empty as Pure Being or Pure Nothing. Still, says Hegel, it has an edge to it, a vibrancy, a sense of potential. It is "an unsteady unrest which sinks into a restful result." (Here I am reminded of the "false vacuum" that, according to current cosmological theory, engendered the Big Bang—another sort of pure becoming.) With some additional teasing and prodding by Hegel, Becoming is made to yield all sorts of still more refined determinations: quantity, quality, and measure, nature and history, art, religion, and philosophy—the whole dialectical process culminating in what he considered to be the perfection of the Prussian State—or what I considered to be the perfection of the Faubourg Saint-Germain *au beau soleil du printemps*.

"So *that's* how all this got here!" I think to myself as I look up from the book.

I may be forgiven for being facetious. Hegel had a gift for eliciting facetiousness in his readers. Wasn't it Bertrand Russell who remarked of Hegel's *Logic*, "The worse your logic, the more interesting the consequences to which it gives rise"? And wasn't it Schopenhauer who derisively credited Hegel with "an ontological proof of absolutely everything"?

What can make Hegel seem so preposterous is the way he equates thought with reality. The world, for him, is ultimately a play of concepts. It is the mind coming to know itself. But what could account for the *existence* of this mind? In what psychic arena, exactly, was Hegel's dialectical orgy supposed to be taking place?

Flipping to the end of *Logic*, I begin to divine the answer. This mind bootstraps itself into existence by constituting its own consciousness. Like Aristotle's God, it is self-thinking thought—only Hegel calls it not "God," but "Absolute Idea."

I come upon Hegel's definition of Absolute Idea: "The Idea, as unity of the Subjective and Objective Idea, is the notion of the Idea—a notion whose object is the Idea as such, and for which the objective is Idea—an Object which embraces all characteristics in its unity."

Russell called this definition "very obscure." I think he was being charitable. Hegel's rhetorical fogginess did not deter French philosophers like Sartre and Merleau-Ponty. They reveled in the air of profundity it gave his dialectic, and emulated it in their own works. For them, Hegel was a model for how an intellectual could "possess the world," as Sartre put it, by thinking alone.

Today, French thinkers still imbibe Hegel with their mother's milk—or, at the very latest, as teenagers at the *lycée*. And here I am, an American weaned on logic of a drier sort, in a state of intellectual prostration after spending just a couple of hours wrestling with his dialectic. Maybe, I think to myself, it is time once again to leave the intellectually inspissate atmosphere of Paris for the clearer metaphysical air of the British Isles.

Or maybe I'm just suffering from the effects of excessive caffeine intake. As a restorative, I decide to order a nice tall glass of my favorite brand of Scotch whiskey—neat. After some minutes, I succeed in attracting the waiter's attention.

"Un grand verre de Glenfiddich, s'il vous plaît," I say. *"Sans glace."*

"Glen-FEE-DEESH," the waiter replies unsmilingly, presuming to correct my pronunciation.

It's definitely time to leave Paris.

12

THE LAST WORD FROM ALL SOULS

No question is more sublime than why there is a Universe: why
there is anything rather than nothing.

—DEREK PARFIT

I always knew that my quest for a resolution to the mystery of Being would bring me back to Oxford. And here I was, standing on the threshold of its most ethereal redoubt, the College of All Souls. I felt a little like Dorothy at the doors to the Emerald City. Inside was a wizard who might well have the final word on the question *Why is there something rather than nothing?* I hoped he would vouchsafe that word to me. And so he did, after a fashion. What I hadn't counted on was that I would get a free lunch in the bargain.

ON MY WAY from Paris back to Oxford, I had stopped for a couple of days in London—not for diversion, but to do some serious swotting up. I had made arrangements to stay at the Athenaeum Club, on Pall Mall. I arrived on a Saturday. The club was closed for the weekend. But when I rang the bell, a porter appeared at the door and let me in. He took me through the crepuscular entrance hall and past the grand staircase, above which hung a large clock. When I looked up to see the time, I noticed that the clock had two numerals for seven, but none for eight. Why, I wondered aloud, was that?

"No one really knows, sir," the porter said, possibly with a wink.

Mystère.

At the rear of the entrance hall was an old and tiny elevator. We took it all the way up to the club's attic floor. I was then conducted through a maze of narrow hallways to what would be my bedroom. It was on the small side, with a couple of little windows that looked out over the statue of Pallas Athene above the portico of the club onto Waterloo Place. Adjoined to the bedroom, happily, was a capacious bath, with a large old-fashioned tub in the middle.

The Athenaeum Club has an impressive library, but I had brought my own reading material to London. It consisted of a Trollope novel—several scenes of which, as it happens, took place on the Doric-columned portico of this very club—and an essay, clipped from an old issue of the *London Review of Books*, by an English philosopher named Derek Parfit. The essay's title was "Why Anything? Why This?"

My familiarity with Parfit as a thinker of rare originality stretched back to my undergraduate days. One summer vacation, while backpacking across Europe, I happened to be carrying around with me a little paperback anthology on the philosophy of mind. The last paper in that anthology, titled "Personal Identity," was by Parfit, and I'll never forget how, when I finally got around to reading it on a long train ride from Salzburg to Venice, it shook up my own sense of self. (Nor will I ever forgot how the prodigious quantity of bread, cheese, and dried sausage I devoured over the course of that train ride fortified my sense of corporeality.) Through a brisk and brilliant series of thought experiments, involving the successive fissioning and fusing of different selves, Parfit arrived at a conclusion that would have astonished even Proust: *personal identity is not what matters.* The permanent, identical "I" is a fiction, not a fact. There may be no determinate answer to whether, say, the callow JH who read Parfit's essay as a student is the same self as the autumnal JH who is typing these words now.

That was how Parfit first came to my awareness. Some years later, in 1984 (by which time I was a philosophy grad student at Columbia University), he published a big book called *Reasons and Persons.* Here he meticulously drew out the implications of his theory of personal identity for morality and rationality, for our obligations to future generations and our attitude toward death. Many of Parfit's conclusions—that we are not what we believe ourselves to be; that it is often rational to act against our self-interest; that our standard morality is logically self-defeating—were disquieting, to say the least. "The truth is very different from what we are inclined to believe," the author coolly declared. But so lucid and powerful were Parfit's arguments that the book gave rise to a veritable cottage industry of commentary in the English-speaking philosophical world. Now Parfit had turned his attention to the question that had engrossed me, the question that he himself considered to be the most "sublime" of all: why is there something rather than nothing? And he had managed to coax his thoughts on the matter into a spare, if sometimes gnomic, essay—one that I knew I had better master before meeting him.

And I *was* going to meet him. "I'm still very interested in 'Why there is something rather than nothing,'" Parfit had responded when I wrote to him a few months earlier. As to the interview I proposed, he wrote, "I'm sure I'd enjoy it." However, he added that since he was very slow in formulating his thoughts, he would prefer not to be quoted verbatim. Instead, he would try to answer any questions I had about his written work with a "yes" or "no" or some other brief response.

I spent much of that weekend in the bathtub under the roof of the Athenaeum, contentedly reading, soaking, sipping claret obligingly brought up to me by the porter from the club's cellar, and pondering. Winston Churchill would have approved.

THERE ARE TWO broad questions we can ask about the world: *why* it is, and *how* it is. Most of the thinkers I had encountered so far believed that the *why* question should come first. Once you know *why* the world is, they maintained, you'll have a pretty good notion of *how* it is. Suppose, like John Leslie, or Plato and Leibniz before him, you believe that the world exists because it *ought* to exist. Then you would expect the world to be a very good world. And if the part of it you observed didn't look especially good, you would conclude—again, like Leslie—that it must be just a tiny bit of a larger reality that, on the whole, *was* very good—*infinitely* good, in fact.

So one way of reasoning about the world is to go from *why* to *how*. But another, less obvious way is to move in the opposite direction. Suppose you look around at the world and notice that it has some special feature, one that marks it off from all of the other ways reality might have been. Perhaps, you might think, this peculiarity in *how* the world is could furnish a clue as to *why* the world is.

Going from *how* to *why* was, I discovered, the essence of Parfit's approach. And his reversal of the usual explanatory vector made me see the mystery of existence in an entirely new light.

Think, Parfit begins, of all the different ways that reality could have turned out. One possible way, of course, is our own world—the universe that came into existence 14 billion years ago with the Big Bang. But reality might encompass more than just our world. There could be other worlds that exist in parallel with our own, even if we do not have direct access to them. And these worlds might be different in important respects from our world—in their histories, in their governing laws (or lack thereof), and in the nature of the substances that constitute them. Each of these individual worlds is what Parfit calls a "local" possibility. And the entire ensemble of individual worlds that might exist together adds up to a "cosmic" possibility.

"*Cosmic* possibilities," Parfit says, "cover everything that ever exists, and are the different ways that the whole of reality might be. Only one such possibility can be actual, or the one that obtains. *Local* possibilities are the different ways that some part of reality, or local world, might be. If some local world exists, that leaves it open whether other worlds exist."

So what sort of cosmic possibilities are there? Well, one cosmic possibility is that *every conceivable world exists*. Parfit calls this fullest of all realities the "All Worlds" possibility. At the other extreme is the cosmic possibility in which *no worlds at all exist*. This Parfit calls the "Null" possibility. In between the All Worlds possibility and the Null possibility range an infinity of intermediate cosmic possibilities. One of them is the possibility that all and only *good* worlds exist—that is, all worlds that are on the whole ethically better than nothing. This would be John Leslie's "Axiarchic" possibility. Another cosmic possibility is the one in which our world and 57 other worlds similar but slightly different from it exist. One might call this the "58 Worlds" possibility. Another is that only worlds that conform to a certain set of physical laws exist—say, the laws of string theory. According to the current version of string theory, such worlds number on the order of 10 to the five-hundredth power, making up a cosmic ensemble that physicists call "the Landscape." Still another cosmic possibility is that only those worlds that are devoid of consciousness exist. One might call this the "Zombie" possibility. And still another is that there are exactly seven worlds, each of a single color: respectively red, orange, yellow, green, blue, indigo, and violet. One might call this the "Spectrum" possibility.

The full range of such cosmic possibilities represents every way reality could have turned out. Even sheer nothingness is counted, in the form of the Null possibility. (Logical *im*possibilities, on the other hand, are *not* counted; no cosmic possibility includes a world of square circles or married bachelors.) And of all the possible ways reality could have turned out, one of them *has* to obtain.

Which raises two questions. Which of them *does* obtain? And why?

"These questions are connected," Parfit goes on to observe. "If some possibility would be easier to explain, we have more reason to believe that this possibility obtains."

The *least* puzzling of all of the cosmic possibilities, it would seem, is the Null possibility—that there is nothing at all. This is much the simplest possible reality, as Leibniz pointed out. It is also the only one that does not stand in need of a causal explanation. If there were no worlds at all, then no question would be raised as to what thing or force could have ushered those worlds into being.

But the Null possibility is evidently not the form reality chose to take. "In some way or another," Parfit observes, "a Universe has managed to exist."

And which is the least puzzling cosmic possibility that is consistent with the fact that our universe exists? That would be the All Worlds possibility: that *all* possible universes exist. "With every other cosmic possibility," Parfit writes, "we have a further question. If ours is the only world, we can ask: 'Out of all the possible worlds, why is this the one that exists?' On any version of the Many Worlds Hypothesis, we have a similar question: 'Why do just these worlds exist, with these elements and laws?' But, if *all* these worlds exist, there is no such further question."

The All Worlds possibility is thus the least arbitrary of the cosmic possibilities, since no local possibility is excluded. And this fullest of all possibilities could, for all we know, be the form that reality actually does take.

But what about the other cosmic possibilities? Well, if our world has a net goodness rating above zero, it might be part of the Axiarchic ensemble of worlds, whose existence would be the ethically best. Or if the laws governing our world turn out, in the final theory envisaged by Steven Weinberg, to be exceptionally elegant, then our world might be part of the most beautiful cosmic possibility. Or, if Schopenhauer and Woody Allen are right, our world might well be part of the *worst* cosmic possibility.

The point is that each of these cosmic possibilities has a special feature. The Null is the simplest, the All Worlds is the fullest, the Axiarchic is the best, and so on. Now, suppose that the cosmic possibility that actually obtains is also one that possesses such a special feature. *Perhaps that is no coincidence.* Perhaps that possibility obtains *because* it has this special feature. If that is the case, this special feature in effect *chooses* what reality is like. It is what Parfit calls "the Selector."

Not every special feature a reality might have makes for a credible Selector. Suppose, for example, the 58 Worlds possibility mentioned earlier was how reality turned out. Now, the number 58 does have a special property: it is the smallest number that is the sum of seven different primes $(2 + 3 + 5 + 7 + 11 + 13 + 17 = 58)$. But no one would imagine that this property could explain why reality turned out the way it did. It would be more reasonable to assume that the number of worlds merely happened to be 58. But features like *best, fullest, simplest, most beautiful*, and *least arbitrary* are different. If the cosmic possibility tapped to be reality had one of these features, it would be hard to think of this as simply a matter of chance. More likely, the cosmic possibility became reality *because* it had this feature.

But isn't this use of "because" somewhat mysterious? Of course it is, Parfit admits. But even ordinary causation, he points out, is mysterious. Besides, he says, "if there is some explanation of the whole of reality, we should not expect this explanation to fit neatly into some familiar category. This extraordinary question may have an extraordinary answer."

What Parfit had managed to do, I realized, was to reframe the mystery of existence in a way that made it vastly less mysterious. While everyone else was trying to bridge the unbridgeable gap between being and nothingness, he was running an ontological lottery. Or was it more like a beauty contest—the Miss Cosmos Pageant? The field of contestants comprised all the different ways reality could have turned out—all the cosmic possibilities. And since reality has to be some way or another, one of these cosmic possibilities is bound to prevail, as a matter of logical necessity. There is no conceivable alternative, and hence no need for any sort of "hidden machinery" to ensure that a selection is made. So the Selector, in tipping the outcome, doesn't exert any force or do any actual work.

But what, I wondered, if there is no Selector?

AFTER MY SOLITARY weekend of reading, brooding, soaking, and dozing, it was good to come down to the commodious dining room of the Athenaeum Club on Monday morning and see a couple of dozen young City of London types, nicely turned out in Savile Row bespoke suits and Turnbull & Asser shirts, at breakfast. It reminded me that there are other things (if not necessarily more important things) beyond all this metaphysical fiddle. I picked up a copy of the *Daily Telegraph*, sat down at a table by myself, and ordered a big greasy English breakfast of eggs and kippers and stewed tomatoes. Delicious. A couple of hours later, feeling more sated than I usually did at that time of day, I was boarding an Oxford-bound train at Paddington Station.

En route to Oxford, I continued to think about what the Selector for our world could possibly be. Clearly, it was not simplicity. For, if it had been, the outcome of the reality contest would surely have been the Null Possibility. And whatever else the west London suburbs and commercial areas my train was passing through at the moment might be—drab, dingy, dispiriting—they were not nothing.

As for Platonic goodness being the Selector, as John Leslie believed, I had long ago put that rather too sanguine notion behind me. So, by the way, had Parfit. "We may doubt that our world could be even the least good part of the best possible Universe," he dismissively observed.

But if this world fails to be ethically distinguished, it does seem to be special in other ways. It displays orderly causal patterns. Moreover, the laws that govern it appear to be, on the deepest level, remarkably simple— so simple that, if Steven Weinberg is right, human scientists are today on the verge of discovering them. Surely these two features—causal orderliness and nomological simplicity—mark off the actual world from the great ruck of messy and complicated cosmic possibilities.

This sort of thinking had led Parfit to the tentative conclusion that there might be at least two "partial Selectors" for reality: being governed by laws, and possessing simple laws. And could there be still others that we have not yet noticed? Possibly. "But observation can take us only part of the way," he observed. "If we can get further, that will have to be by pure reasoning." Such reasoning aims at the highest principle governing reality—the same principle that physicists are trying to discover. Thus, said Parfit, "there is no clear boundary here between philosophy and science."

Hello! The train is pulling into Oxford already, right at the prick of noon.

FROM THE TRAIN station it was just a short walk to the town center—a walk with which I was by now well familiar. "Come to All Souls College in the High Street at 1 p.m. and ask the Porter to call me from the lodge by the College gate," Parfit had instructed me in his letter.

Since I had a little time to kill, I dropped into Blackwell's on Broad Street, the best scholarly bookshop in the English-speaking world. I headed downstairs to the vast philosophy section, where, after browsing a bit, I found a wonderful book of photo-portraits of the greatest living philosophers, taken by a photographer named Steve Pyke. Parfit was among the subjects. His appearance was certainly striking: an elongated face, featured with thin lips, a granite nose, and wide pensive eyes, was surmounted by a luxuriant profusion of curly silver-white hair, which extended down the sides of his head almost to the level of his chin. Each photo was captioned with a personal statement by the philosopher who had posed for it. Parfit's read, "What interests me most are the metaphysical questions whose answers can affect our emotions and have rational and moral significance. Why does the Universe exist? What makes us the same person throughout our lives? Do we have free will? Is time's passage an illusion?"

A quarter of an hour later, I was peering through the rather forbidding gate of All Souls. THE COLLEGE IS CLOSED, announced one sign. QUIET PLEASE, said another. Beyond the gate, I could see a courtyard with two manicured rectangles of grass.

I made myself known to the college porter, who was dour of aspect, and waited as he rang up my host-to-be.

All Souls is a storied place. ("All Souls, no bodies," says the wag.) One occasional visitor to All Souls when he was an Oxford undergraduate in the 1960s was Christopher Hitchens, who described it as "a florid antique shop that admitted no students and guarded only the exalted privileges of its 'fellows,' a den of iniquity to every egalitarian and a place where silver candelabras and goblets adorned a nightly debauch of venison and port." The fellows of All Souls, seventy-six in number, are selected from the most august ranks of the British academy and public life. Having no tutoring duties, they are free to pursue, amid sumptuous surroundings, a life of pure scholarship and speculative thought—relieved, perhaps, by internal politics and gossip. Parfit, somewhat unusually, had spent the whole of his career there, having been elected a "prize fellow" in 1967, fresh out of his undergraduate days at Balliol College.

And here he was, bounding toward me diagonally across the quadrangle—a tall, gangling, smiling fellow, whose unruly mop of argent tresses fulfilled the promise of the photo I had just seen. He was wearing a bright-red tie, which rhymed with his rather rubicund face. We shook hands and exchanged greetings. I offered to take him to one of the better restaurants out on the High Street for a long wine-soaked lunch.

"No," he said, "I'm giving *you* lunch."

He led me inside the college. "This is the best view in all of Oxford," he said, gesturing out a large window toward Radcliffe Camera, the old library of Oxford. "The dome is by Hawksmoor!"

I remembered having heard that Parfit was a keen architectural photographer.

Lunch was being served to the fellows of All Souls in "the Buttery," a Gothic dining room with a lofty coffered ceiling and highly resonant acoustics. Parfit invited me to help myself at the buffet, where I filled my plate with avocado salad and bread. We sat down to eat and talk.

Parfit told me about his life. He had been very pious as a young child, he said, but he gave up religion at the age of eight or nine. He remembered, when looking at pictures of the crucifixion, how he felt the most pity for the bad thief—"because, unlike Jesus and the good thief, he's going to *hell* after he suffers and dies on the cross."

Then he talked about mathematics, at which, he said, he was terrible. He expressed amazement that mathematics could be so complicated. A mathematician had told him that 80 percent of mathematics was about infinity. And he was horrified to learn that there was more than one infinity!

Even though his father wanted him to be a scientist, Parfit continued, he decided that he would become a philosopher. He hated the "scientizing" of philosophy, the main influences behind which, he felt, were Quine and Wittgenstein. He also hated the "naturalizing" of epistemology—the idea that the project of justifying our knowledge should be taken away from philosophers and given to cognitive scientists.

Then the talk turned to moral philosophy, which, he told me, was his main interest at the moment. Unlike many moral philosophers these days, he said, he believed that we have objective reasons to be moral, reasons that do not depend on our inclinations—adding that he would be "embarrassed even to have to defend that claim before a non-university audience." He was appalled, he said, at some of the crazy views that contemporary philosophers had argued for, like the view that only desires can give rise to reasons.

Parfit winced, as if in pain, when mentioning such distasteful views, and often flung his arms toward the coffered ceiling in exasperation. He was equally animated when putting forth the views that he favored, leaning close to me, grinning, and vigorously nodding.

When lunch was finished, we retired into an adjoining parlor to have coffee by the fireplace and talk about why there is Something rather than Nothing.

PARFIT, AS I mentioned earlier, had declined to be quoted at length on the matter. He did, however, say that he would answer my questions with a brief affirmative or negative reply. And I had two main questions, one easy and one hard.

The easy one had to do with nothingness. Parfit clearly believed that nothingness was a logically coherent idea. Indeed, he thought it was one of the ways reality could have turned out. "It might have been true," he had written, "that nothing ever existed: no minds, no atoms, no space, no time." Nothingness was therefore included among his cosmic possibilities, in the form of the Null possibility.

But was nothingness also a *local* possibility? That is, could it coexist with a world of being?

The philosopher Robert Nozick, for one, had thought that it could. If reality was as full as possible, encompassing every conceivable world, then one of those worlds would perforce consist of absolutely nothing. That, at least, was what Nozick believed. So the question *Why is there something rather than nothing?* on his way of thinking, might have a simple answer: There isn't. There's both.

Nozick's reasoning has convinced some scientists, including his onetime Harvard student, the string theorist Brian Greene. "In the Ultimate Multiverse," Greene has written, "a universe consisting of nothing *does* exist." Again, reality embraces both something *and* nothing.

And, from a somewhat different angle, Jean-Paul Sartre agreed, declaring that "Nothingness haunts Being."

But the notion that reality could embrace both being and nothingness struck me as wrong-headed, and I said so to Parfit. How could it make sense to talk of adjoining a "world of nothing" to an ensemble of something-worlds? It would not be like adding a barren planet, or a region of empty space. For a barren planet is something. And so, pretty much everyone agrees, is a region of empty space. Space has features. It can, for example, be either finite or infinite in extent. Nothingness is not like that.

I wanted to put the point in the form of an equation:

$$Something + Nothing = Something$$

But even that seemed too weak. To add "nothing" to a cosmic possibility was an empty gesture. It was to do nothing at all.

Parfit agreed. Nozick and the others were wrong, he believed. Nothingness is not a local possibility; it cannot be one world among many. The only reality consistent with Nothingness is the reality consisting of *no* worlds at all: the Null Possibility. You can have two different somethings, but you can't have both something and nothing. It's strictly an either/or deal.

My second question for Parfit was a deeper one. Suppose he was right in thinking that what he called a Selector might yield the explanation for why reality took the particular form that it did. Would that be the end of the matter? Does cosmic explanation stop at the Selector level? Or could there be a further explanation as to why some particular Selector, among all the other plausible rival Selectors, prevailed?

Think again of the analogy to the Miss Cosmos Pageant. The contestants are all the conceivable ways reality might have been—all the cosmic possibilities. One of these contestants has to be crowned the winner. Suppose the winner turns out to be the ethically best cosmic possibility: Miss Infinitely Good. Then we might suspect that the judges used goodness as the Selector: that, after all, would explain the choice of Miss Infinitely Good as the winner. But couldn't we go on to ask *why* the judges used goodness as their Selector rather than, say, simplicity, elegance, or fullness?

On the other hand, suppose the winner of the Miss Cosmos Pageant turned out to have no special features. Suppose it was Miss Mediocre. Then we might conclude that the judges didn't use any Selector at all. They didn't care about what the contestants were like, what their special virtues might be. They simply drew straws. But couldn't we go on to ask *why* the pageant's judges didn't bother to use a Selector to choose the winner?

Parfit acknowledged the need for further cosmic explanation. "Reality may happen to be as it is, or there may be some Selector," he had written. "Whichever of these is true, it may happen to be true, or there may be some higher Selector. These are the different possibilities at the next explanatory level, so we are back with our two questions: which obtains, and why?"

So first you need a Selector to explain why reality is the way it is. Then you need a meta-Selector at the next explanatory level to account for why *that* Selector was the operative one in choosing how the world turned out. And then you need a meta-meta-Selector at a still higher explanatory level to account for why *that* meta-Selector was tapped. And so on. Could this explanatory regress ever come to an end? And, if so, *how* could it end? With some highest Selector? Then wouldn't that be the ultimate brute fact?

When I put this question to Parfit, he conceded that the quest to explain reality would likely end with such a brute fact. How could this be avoided? You might try to avoid it by saying that a Selector could select itself. For instance, if goodness proved to be the highest Selector, one might try to say that this is true because it's for the best. That is, goodness chose itself as the ruler of reality. But Parfit didn't buy that. "Just as God could not make himself exist, no Selector could make itself the one that, at the highest level, rules," he maintained. "No Selector could settle whether it rules, since it cannot settle anything unless it *does* rule."

Nonetheless, Parfit insisted, an explanation that ends with a brute fact is better than no explanation at all. Indeed, he observed, scientific explanations invariably take this form. Such an explanation can still help us to discover what reality, on its grandest scale, might actually be like—say, by giving us a reason to believe that reality comprises worlds beyond our own.

As Parfit sipped his coffee, I brought out a little diagram that I had made over the weekend. It showed how the various Selectors might be related to one another, and to reality. At the bottom of the paper, I had sketched the reality level. There I had set out some of the cosmic possibilities that Parfit had talked about. At the level above that—the first explanatory level—I had jotted down some of the plausible Selectors. And at the level above that— the second explanatory level—I had indicated some of the meta-Selectors. Then I had drawn arrows between the different levels to indicate the various explanatory relationships that might obtain. The diagram looked like the one on the facing page.

"I see you've worked out all the logical implications," Parfit said as he leaned forward and squinted at my diagram.

Most of these implications had already been drawn by Parfit himself, and they were pretty straightforward. The Simplicity Selector, for example, picks out Null possibility from among the cosmic possibilities. Thus, if there had been nothing at all, that would have been explained by the fact that nothingness was the simplest way reality might have been.

Similarly, the Goodness Selector picked out the Axiarchic possibility—a universe consisting only of good worlds. Thus, if reality turns out to take that form, it would be explained by the fact that this was the best way reality could have been. But, if reality did take that form, what could explain the fact that the Goodness Selector ruled? Only that the Goodness Selector, being so good, was itself selected by Goodness at the meta-level. And here, as Parfit had observed, we run into a problem: A Selector cannot select itself. It cannot settle whether it rules unless it *does* rule. Otherwise put, no explanation of reality is capable of explaining itself.

To indicate that Goodness could not, on pain of circularity, explain itself, I had drawn an "X" across the arrow leading from Goodness at the meta-Selector level to Goodness at the Selector level.

But not all Selectors are prone to this sort of circularity. That is, not all Selectors select themselves. And that fact was reflected in what I felt was the most interesting arrow in my diagram: the one that went from Simplicity at the meta-explanatory level to Null at the explanatory level.

This arrow, too, was inspired by what Parfit had written. At the very end of his "Why Anything?" essay, he had made an alluring observation: "just as the simplest cosmic possibility is that nothing ever exists, the simplest explanatory possibility is that there is no Selector." I had taken this to mean that the No Selector possibility at the explanatory level is like the Null possibility at the reality level: each would be explained by Simplicity. Then if Simplicity rules at the meta-explanatory level, it would not pick itself as the Selector at the explanatory level. Rather, it would decree that there would be no Selector at all.

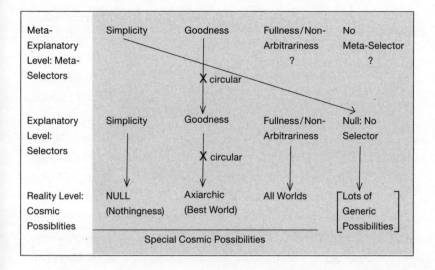

Was this indeed what Parfit meant?

"That's right," he said with a smile.

And what would reality look like if there were *no* Selector? Well, it almost certainly *wouldn't* take the very special form of nothingness, the emptiest of all cosmic possibilities. "If there is no Selector," Parfit had written, "we should not expect that there would also be no Universe. That would be an extreme coincidence." By the same token, it seemed to me, we should not expect it to take *any other* special form. If there were no Selector, we should not expect reality to be as full as it could possibly be, as good as it could possibly be, as bad as it could possibly be, as mathematically neat as it could possibly be, and so on. Rather, we should expect such a blindly chosen reality to be one of the countless cosmic possibilities that have *no special feature at all*. In other words, we should expect reality to be *thoroughly mediocre*. Did Parfit agree with this reasoning?

He nodded that he did.

So if Simplicity *is* the ultimate Selector, this would explain why there is something rather than nothing! Heidegger, in his fuddled way, may have had a point after all. *Das Nichts selbst nichtet*: "Nothing noths itself." If nothingness prevails at the explanatory level, then there is no Selector explaining the way reality turned out. But if there is no Selector, then the way reality turned out would be a *random* matter. In that case, it would be exceedingly odd if reality turned out to be nothingness. For the Null Possibility is a very special outcome, being the simplest of all cosmic possibilities. So nothing (at the explanatory level) noths itself (at the cosmic level)—with the upshot that reality comprises something rather than nothing. All by dint of Simplicity ruling at the highest level.

If Simplicity is the ultimate explanation of things, this would also account for why the actual cosmos seems to be so disappointingly average: an indifferent mixture of good and evil, of beauty and ugliness, of causal order and random chaos; inconceivably vast, yet falling well short of the full cornucopia of possible being. Reality is neither a pristine Nothing nor an all-fecund Everything. It's a cosmic junk shot.

Such was the conclusion that I had teased out of Parfit's scheme. But, frustratingly, it still fell short of being a *complete* explanation. If Simplicity did indeed rule at the highest level, did this merely *happen* to be true? What about rival meta-Selectors, like Fullness? (I had put a question mark under it on my diagram.) And what if there were no meta-Selector at all? (Another question mark on my diagram.) Was the most general explanation of reality doomed to end with an inexplicable brute fact?

Parfit had done his bit. He had dispelled much of the fog surrounding the mystery of existence. And he had given me a very nice lunch in the bargain. It was time for him to return to his study, where he would reimmerse himself in questions of moral philosophy, of values and desires and reasons. And it was time for me to leave the rarefied cloister of All Souls and return to the rude world of Vile Bodies.

I thanked Parfit effusively, saw myself to the college gate, and turned onto the High Street, whose shadows were lengthened by the late-afternoon sun.

A WEEK LATER I was back in New York, still vexing over the now-crumpled diagram I had shown to Parfit. Then, walking one evening in the tonic squalor of the East Village, a million miles from All Souls, I had an epiphany. The last piece of logic snapped into place. I had the proof.

Epistolary Interlude

The Proof

Wednesday morning
2 Fifth Avenue, New York

Dear Professor Parfit,

It was so nice spending the afternoon with you at All Souls.
In reflecting on our conversation, I think I may have hit upon a
complete and unique explanation for the most general form reality takes—an explanation that finally resolves the question, *Why
is there something rather than nothing?*

I begin by assuming two principles:

(I) For every truth, there is an explanation of why it is true.
(II) No truth explains itself.

The first of these, of course, is what Leibniz called the Principle of Sufficient Reason. It says that there are no brute facts.
I take Sufficient Reason to be not so much a truth in itself, but
rather a provisional guide to inquiry, one that says, "Always look
for an explanation unless you find yourself in a situation where
further explanation is impossible."

The second principle is a more general version of your point
that no Selector can select itself. It is meant to rule out circularity. A cause cannot cause itself. A rationale cannot justify itself.
God cannot create himself. A set cannot be a member of itself.

In set theory, this is called the Axiom of Foundation. So I'll call principle (II) "Foundation."

Now here's the argument that there is one, and only one, complete explanation for the form that reality takes.

At level 0, the level of reality, you have all the "cosmic possibilities" for how reality might have turned out. These range from the Null possibility to the All Worlds possibility, and they include every one of the countless intermediate possibilities, where some conceivable worlds of some types exist, but not others. One of these cosmic possibilities *has* to obtain, as a matter of logical necessity. Call the cosmic possibility that actually does obtain *A*, for "actual."

At level 1, the lowest explanatory level, you have all the plausible Selectors—all of the possible explanations that might account for the way reality at level 0 turns out. These include Simplicity, Goodness, Causal Orderliness, and Fullness, as well as the No Selector possibility—the possibility that there is no explanation at all.

At level 2, the meta-explanatory level, you have all the plausible meta-Selectors—all of the possible explanations that might account for which Selector prevails at level 1. These include—again—Simplicity, Goodness, Causal Orderliness, and Fullness, as well as the No Meta-Selector possibility.

Now let's consider some cases.

First, suppose that no Selector explains why reality takes the form it does, and that there is no further explanation for *why* there is no Selector. Then it is a brute fact that reality takes the form *A*. But this violates Sufficient Reason. Dead end.

Next, suppose that one of the Selectors at level 1 *does* explain why reality takes form *A*. Call this Selector *S*. Then either there is an explanation for why *S* prevailed in determining reality, or there isn't. If there isn't, then it is a brute fact that *S* is the Selector. But this violates Sufficient Reason. Dead end.

So suppose there *is* an explanation for *S* being the Selector.

In other words, suppose there is a meta-Selector (at level 2) that selected S (at level 1). Call this meta-Selector M.

Now ask, What could M be?

We know that M could not be the same as S. That would violate the Principle of Foundation. For instance, if S were Goodness (in which case reality would have taken the ethically best possible form), the explanation for that could not be that it is *ethically best* that Goodness should be the Selector. The same goes for the other Selectors that pick out cosmic possibilities intermediate between the Null possibility and the All Worlds possibility—like the Causal Orderliness Selector, or the Mathematical Elegance Selector, or the Evil Selector. These Selectors all select themselves at the meta-level, and that is circular.

In fact, only *two* meta-Selectors at level 2 could serve as M. These are Simplicity and Fullness. Neither of these selects itself, and hence neither violates the Principle of Foundation. If Simplicity were the meta-Selector that prevailed at level 2, it would not select itself at level 1. Rather, it would select the No Selector possibility, since that is the simplest of the explanatory possibilities—that there is no explanation. And if Fullness were the meta-Selector that prevailed at level 2, it would not select itself at level 1. Rather, it would select *all* the Selectors at level 1.

Thus, assuming the Principle of Foundation, it is a logical truth that there are only two possible meta-Selectors at level 2: Simplicity and Fullness. One or the other of them *has* to constitute the ultimate explanation.

So there are two cases left to consider.

Case 1: *Simplicity is the meta-Selector.* Then it would pick out the No Selector possibility at level 1 (just as Simplicity at level 1 would pick out the Null possibility at level 0). But if there is no Selector at level 1, then A, the cosmic possibility that reality takes, would be randomly picked, a matter of pure chance. Yet this would not be a brute fact; rather, it would be explained by Simplicity at the meta-explanatory level.

Case 2: *Fullness is the meta-Selector.* Then it would pick out *all* the Selectors at level 1 (just as Fullness at level 1 would pick out the All Worlds possibility at level 0). But it is logically impossible for *all* Selectors at level 1 to dictate the form reality takes. That is because they contradict one another. Reality cannot be perfectly full and perfectly empty at the same time. Nor can it be ethically the best and causally the most orderly at the same time (since the occasional miracle could make reality better). And it certainly can't be the ethically best and the most evil at the same time. At most, the Selectors at level 1 could all operate together only as *partial* Selectors. Then A, the cosmic possibility selected at level 0 to be reality, would be *thoroughly mediocre.* It would be as full and as empty as possible, as good and as evil as possible, as orderly and as chaotic as possible, as elegant and as ugly as possible, and so on.

In case 1, A would be chosen at random from among the cosmic possibilities. In case 2, A would be the most mediocre of the cosmic possibilities. These are the only level 0 reality outcomes that are consistent with the principles of Sufficient Reason and Foundation. And they are overwhelmingly likely to amount to the same thing! A cosmic possibility chosen at random is overwhelmingly likely to be thoroughly mediocre.

This is a matter of sheer numbers. Of all the possible forms reality might take, only a vanishingly small proportion of them possess special features—like being perfectly simple, or perfectly good, or perfectly full. The vast majority have no special feature at all. They are *generic* realities.

And what would such a generic reality look like? First of all, it would be *infinite.* Realities consisting of infinitely many worlds are vastly more numerous than those consisting of finitely many worlds. (This, of course, follows from an elementary result in set theory. The number of *finite* subsets of the natural numbers, though infinite itself, is of a smaller order of infinity than the number of *infinite* subsets of the natural numbers.)

But even in its infinity, a generic reality would fall far short of encompassing everything possible—infinitely short, in fact. (In set-theoretic terms, the complement of an infinite generic reality is also infinite.) A generic reality is thus infinitely removed from the All Worlds Possibility as well as from the Null Possibility.

A generic reality, being infinite, would perforce have lots of local regions that appeared to be special in one respect or another. Think of an infinite sequence of random coin tosses: 1 for "heads," 0 for "tails." Even though the sequence as a whole will be patternless, it is guaranteed to contain—by pure chance—all conceivable local patterns. There will be stretches of perfect fullness, consisting of a long run of 1's. There will be stretches of perfect emptiness, consisting of a long run of 0's. There will be stretches that constitute the most beautiful possible patterns, and stretches that constitute the ugliest possible patterns. There will be stretches that appear meaningful, that seem to harbor hidden messages and purposes. But each such local meaning/message/purpose will be contradicted by another local meaning/message/purpose elsewhere in the generic reality. So they will add up to cosmic meaninglessness.

This is the sort of reality that is overwhelmingly likely to result when the meta-Selector is either Simplicity (case 1) or Fullness (case 2). And since these are the *only* logical possibilities consistent with the principles of Sufficient Reason and Foundation, this is the way reality *must be* if these principles are valid. So we have a *complete* explanation for the form that reality takes—no brute facts, no loose ends. It is an explanation that answers both of the questions with which you started your metaphysical inquiry: Why anything? Why this?

What if, on further empirical inspection, reality doesn't turn out to look so generic after all? What if it turns out, rather, to look as ethically good as possible, the way John Leslie believes it must be? Or as plenitudinous as possible, the way Robert Nozick thought it might be? Or what if a God suddenly manifests

himself as the font of being. Then, assuming my logic is correct, either the Principle of Sufficient Reason or the Principle
of Foundation (or both) must be violated. There must be an
ultimate brute fact or a self-caused cause after all. But such an
appearance of cosmic specialness might well be an illusion—one
to which we humans are vulnerable because our imaginative
powers, partaking of the mediocrity of reality as a whole, are too
limited to see this reality as it truly is.

Please don't feel obligated to respond. I know you're very
busy with more important things. And thanks again for lunch!

Gratefully,
Jim Holt

Wednesday evening
All Souls College, Oxford

Dear Jim,
Thanks for this message, which is very interesting. I shall
have to think about it carefully . . .

Best wishes,
Derek

13

THE WORLD AS A BIT OF LIGHT VERSE

Late winter in Manhattan. Afternoon. A siren in the distance. (There is always a siren in the distance.) The phone rings. It's John Updike.

I had been expecting the call. Earlier that month, I had sent a letter to Updike describing my interest in the mystery of existence. I had guessed, I said, that he shared this interest, and I wondered whether he would be willing to talk about the matter. I included my phone number in case he did.

A week later, I received a plain postcard with Updike's return address on the front and a long type-written paragraph crowded onto the back. The occasional typo had been corrected in pen with a proofreader's "delete" or "transpose" sign. At the bottom, in blue ink, it was signed "J.U."

"I'd be happy to talk to you about something rather than nothing," Updike had typed, "with the warning that I have no thoughts." He then, in a trio of brisk sentences, mentioned the dimensionality of reality, the possibility of positive and negative being, and the anthropic principle—the last of which, he cryptically added, "to some extent works for somethingness." Then, as a comment on the mysteriousness of it all, came the kicker:

"Beats me, actually; but who doesn't love the universe?"

THAT UPDIKE LOVED the universe had long been obvious to me. His novels and stories are suffused with the sheer sweetness of being. We "skate upon an intense radiance we do not see because we see nothing else," he wrote in a memoir of his youth. "And in fact there is a color, a quiet but tireless goodness that things at rest, like a brick wall or a small stone, seem to affirm."

In this respect, Updike was the anti–Woody Allen.

But, in another respect, he was at one with Woody Allen. He shared the same horror of eternal nothingness—and the conviction that sex offered a psychological hedge against it. Indeed, he found that his phobia of nonbeing was inversely proportional to his carnal flourishing—a point he put in succinct mathematical form in his 1969 credo poem, "Midpoint":

$$ASS = 1 / ANGST$$

But it was not only eros that fortified Updike against the terrors of nothingness. He also claimed to draw consolation from religion—specifically, from a leap-of-faith version of Christianity—and the hope it offered of all-encompassing grace and personal salvation. Here his heroes were Pascal and Kierkegaard and, especially, Karl Barth. "Barth's theology, at one point in my life, seemed alone to be supporting it (my life)," Updike once observed. He professed to share Barth's belief that God is *totaliter aliter*—wholly other—and that the divine mysteries could not be approached by rational thought. He was also drawn to Barth's somewhat mystical equation of nothingness with evil. In an early collection of writings, *Picked-up Pieces*, Updike darkly dilates on the idea of "Satanic nothingness"—and then, as if in search of metaphysical relief, transitions directly to an essay on golf.

Updike's obsession with sex and death, with the goodness of being and the evil of nonbeing, is perhaps not unusual in the literary profession. But only with Updike do you find the mystery of existence figuring directly and explicitly in his fiction. His 1986 novel, *Roger's Version*, a merry roundelay of theology, science, and sex, culminates in a virtuoso passage that explains, over the course of nearly ten pages, "how things popped up out of nothing." The explanation is delivered in the course of a cocktail party. It is meant to shatter both the faith and the spirit of the character Dale Kohler, a twenty-eight-year-old Jesus freak and software whiz who has had the effrontery to try to prove the existence of God through a computer-run numerical analysis of the Big Bang. Dale also had the effrontery to sleep with the wife of the novel's narrator and title character, a middle-aged professor of divinity named Roger Lambert.

Like Updike himself, the cuckolded Roger is "a Barthian all the way." He resents not only the younger man's priapic poaching of his sexually athletic wife, but also his "obscene cosmological prying" into the realm of the numinous. A God whose existence could be scientifically proved—let alone one who left his fingerprints all over the Big Bang—would not be God, not, at least, the *totaliter aliter* God of Barth. So, at the end of the novel, Dale gets a double comeuppance. Roger himself doles out the theological punishment for Dale's heresy. And he gets a friend, a molecular biologist named Myron Kriegman, to ambush Dale from the scientific side. Kriegman does this by accosting Dale at the aforementioned cocktail party and discomfiting him with arguments purporting to show that the physical universe, without the least need of divine assistance, created itself out of nothing at all.

"As you know, inside the Planck length and the Planck duration you have this spacetime foam where the quantum fluctuations from matter to non-matter really have very little meaning, mathematically speaking. You have a Higgs field tunneling into a quantum fluctuation through the energy barrier in a false-vacuum state, and you get this bubble of broken symmetry that by negative pressure expands exponentially, and in a couple of microseconds you can have something go from next to nothing to the size and mass of the observable present universe. How about a drink? You look pretty dry, standing there."

So Kriegman begins, in his rapid, rasping voice. Then, having supposedly shown how the universe was generated from "almost nothing," he goes on to explain to the nonplussed Dale that this almost nothing arose from *absolutely* nothing.

"Imagine nothing, a total vacuum. But wait! There's something in it! A kind of dust of structureless points . . ." It is this swirling dust of points, he goes on, that by chance becomes "knotted" or "frozen" into a little bit of structured spacetime. "The seed of the universe has come into being," Kriegman says. And once you've got that seed, "ka-*boom!* Big bang is right around the corner."

And whence does the primordial dust of points arise? Out of nothing at all! Point and anti-point peel apart from the void, the way +1 and –1 can peel apart from 0. "Now you have something, you have *two* somethings, where once you had nothing," Kriegman says. An anti-point is just a point moving backward in time.

"The dust of birth gives birth to time, and time gives birth to the dust of points," Kriegman concludes. "Elegant, huh?"

Elegantly *circular*, one wants to say in the speechless Dale's behalf. Time is needed to give being to the primordial dust of points. But it is the pattern assumed by those points that constitutes time!

No doubt Updike didn't mean for us to take these ideas too seriously. They are being mouthed, after all, by a character in a novel, and a somewhat ridiculous character to boot. (Updike told me in his postcard that he borrowed the bulk of them from the British chemist, and outspoken atheist, Peter Atkins. Atkins himself, I discovered, was aware of the circularity in his cosmogonic scheme, wherein time is required to give life to points and points are required to give life to time. He called it the "cosmic bootstrap," and left the matter at that.) Still, Updike had clearly pondered the mystery of being from the scientific as well as the theological angle. And that was reason enough to seek out his thoughts.

UPDIKE WAS CALLING from his longtime home in the town of Ipswich, on the Massachusetts shore an hour north of Boston. In the background I could hear his visiting grandchildren at play. As he spoke, in his characteristically soft and richly modulated voice, I could see him in my mind's eye: the thick thatch of gray hair, the curved beak of a nose, the mottled, psoriatic complexion, the eyes and mouth forming his habitual expression, that of a man, as Martin Amis once put it, "beset by an embarrassment of delicious drolleries."

I began by asking Updike whether the theology of Karl Barth had really sustained him through a difficult time in his life.

"I've certainly said that and it did seem to be true," he said. "I fell upon Barth having exhausted Kierkegaard as a consoler, and having previously resorted to Chesterton. I discovered Barth through a series of addresses and lectures called *The Word of God and the Word of Man*. He didn't attempt to play anybody's game as far as looking at the Gospels as historic documents or anything. He just said, essentially, that this is a faith—take it or leave it. So yes, I did find Barth comforting, and a couple of my early novels—not so early, actually—are sort of Barthian. *Rabbit Run* certainly presents a Barthian point of view, from the standpoint of a Lutheran minister. And in *Roger's Version*, Barthianism is about the only refuge for Roger from all the besieging elements that would deprive one of one's faith—both science, which Dale tries to use on behalf of the theist point of view, and the watering down of theology with liberal values. On the other hand, the book is kind of a critique of Barthianism too, as being, in the end, terribly arid and self-enclosed. Dale at least has the virtue of trying to reconcile his Christianity with science as it presently exists. And the whole book is kind of a love triangle, in that Roger, rightly or wrongly, imagines that his wife is having an affair with young Dale up in her studio. So the conflict between the two men is sort of a tussle over . . . her name escapes me . . ."

"Esther," I interject.

"Right, Esther . . . I like her, she appears in a bumble-bee kind of dress . . . stripes, big broad stripes bracing her hips. So Roger tries to organize this party so that a number of agile-tongued scientists will be there to debunk, piece by piece, Dale's natural theology."

Was their scientific account of the origin of the universe from nothingness meant to be convincing?

"Not entirely, and that's an embarrassment for science. Science aspires, like theology used to, to explain absolutely everything. But how can you cross this enormous gulf between nothing and something? And not just something, a whole universe. So much . . . I mean the universe is very big. Ugh! I mean, it's big beyond imagining squared!"

Updike's voice rose a register in genuine wonderment.

"It's interesting," I said, "that some philosophers are so astonished and awed that anything at all should exist—like Wittgenstein, who said in the *Tractatus* that it's not *how* the world is that is mystical, but *that* it is. And Heidegger, of course, made heavy weather of this too. He claimed that even people who never thought about why there is something rather than nothing were still 'grazed' by the question whether they realized it or not—say, in moments of boredom, when they'd just as soon that nothing at all existed, or in states of joy when everything is transfigured and they see the world anew, as if for the first time. Yet I've run into philosophers who don't see anything very astonishing about existence. And in some moods I agree with them. The question *Why is there something rather than nothing?* sometimes seems vacuous to me. But in other moods it seems very very profound. How does it strike you? Have you ever spent much time brooding over it?"

"Well, to call it 'brooding' would be to dignify it," Updike said. "But I am of the party that thinks that the existence of the world is a kind of miracle. It's the last resort, really, of naturalistic theology. So many other props have been knocked out from under naturalistic theology—the first principle argument that Aristotle set forth, Aquinas's prime mover . . . they're all gone, but the riddle does remain: why is there something instead of nothing? George Steiner is a lesser thinker than Wittgenstein, but I remember him raising this issue. At the last report that reached my ears, Steiner found the existence of a world amazing, and enough of a puzzle to sustain a faith of a sort."

I said, "I had no idea Steiner . . ."

"Yeah, I didn't know he cared either," Updike continued. "And I can't remember where he brought the question up. Steiner has a theological side, which is not evident in everything he writes. But for the scientifically inclined layman, the great hope for explaining 'something out of nothing' is quantum physics, where you have these virtual particles that keep popping up out of the vacuum and disappearing. They're around for miraculously short periods of time, but they're nevertheless indubitably there."

He was careful to pronounce each distinct syllable of "indubitably."

I told Updike that I admired the way he had a character in *Roger's*

Version explain how the universe might have arisen from nothingness via a quantum-mechanical fluctuation. In the decades since he wrote the book, I added, physicists had come up with some very neat scenarios that would allow something to emerge spontaneously out of nothing in accordance with quantum laws. But then, of course, you're faced with the mystery: Where are these laws written? And what gives them the power to command the void?

"Also, the laws amount to a funny way of saying, 'Nothing equals something,'" Updike said, bursting into laughter. "QED! One opinion I've encountered is that, since getting from nothing to something involves time, and time didn't exist before there was something, the whole question is a meaningless one that we should stop asking ourselves. It's beyond our intellectual limits as a species. Put yourself into the position of a dog. A dog is responsive, shows intuition, looks at us with eyes behind which there is intelligence of a sort, and yet a dog must not understand most of the things it sees people doing. It must have no idea how they invented, say, the internal-combustion engine. So maybe what we need to do is imagine that we're dogs and that there are realms that go beyond our understanding. I'm not sure I buy that view, but it is a way of saying that the mystery of being is a *permanent* mystery, at least given the present state of the human brain. I have trouble even believing—and this will offend you—the standard scientific explanation of how the universe rapidly grew from *nearly* nothing. Just think of it. The notion that this planet and all the stars we see, and many thousands of times more than those we see—that all this was once bounded in a point with the size of, what, a period or a grape? How, I ask myself, could that possibly be? And, that said, I sort of move on."

Updike chuckled softly. His mood appeared to lighten.

"The whole idea of inflationary expansion," he continued, "seems sort of put forward on a smile and a shoeshine. Granted, it solves a number of cosmological problems that were embarrassing . . ."

Wait—a smile and a *what*?

"A smile and a shoeshine . . ."

I'd never heard that expression, I said. It's charming.

"Oh, that's what Willy Loman in *Death of a Salesman* was out on. He was out there, as they say at his funeral service, 'on a smile and a shoeshine.' You've never heard that?"

I confessed that I was a theater philistine.

"It's a phrase I can't shake, because, in a way, a writer too is out there on a smile and a shoeshine. Although people maybe don't shine their shoes as much now. It's hard to shine a running shoe."

I always feel virtuous, I told Updike, when I shine my shoes.

"So anyway," he continued, "when you think about it, we rationalists—and we're all, to an extent, rationalist—we accept propositions about the early universe which boggle the mind more than any of the biblical miracles do. Your mind can intuitively grasp the notion of a dead man coming back again to life, as people in deep comas do, and as we do when we wake up every morning out of a sound sleep. But to believe that the universe, immeasurably vast as it appears to be, was once compressed into a tiny space—into a tiny *point*—is in truth very hard to believe. I'm not saying I can disprove the equations that back it up. I'm just saying that it's as much a matter of faith to accept that."

Here I was moved to demur. The theories that imply this picture of the early universe—general relativity, the standard model of particle physics, and so forth—work beautifully at predicting our present-day observations. Even the theory of cosmic inflation, which admittedly is a bit conjectural, has been confirmed by the shape of the cosmic background radiation, as measured by the Hubble space telescope. If these theories are so good at accounting for the evidence we see at present, why shouldn't we trust them as we extrapolate backward in time toward the beginning of the universe?

"I'm just saying I can't trust them," Updike replied. "My reptile brain won't let me. It's impossible to imagine that even the Earth was once compressed to the size of a pea, let alone the whole universe."

Some things that are impossible to imagine, I pointed out, are quite easy to describe mathematically.

"Still," Updike said, warming to the argument, "there have been other intricate systems in the history of mankind. The scholastics in the Middle Ages had a lot of intricacy in their intellectual constructions, and even the Ptolemaic epicycles or whatever were . . . Well, all of this showed a lot of intelligence, and theoretical consistency even, but in the end they collapsed. But, as you say, the evidence piles up. It's been decades and decades since the standard model of physics was proposed, and it checks out to the twelfth decimal point. But this whole string theory business . . . There's never any evidence, just mathematical formulas, right? There are men spending their whole careers working on a theory of something that might not even exist."

Even so, I said, they're doing some beautiful pure mathematics in the process.

"Beautiful in a vacuum!" Updike exclaimed. "What's beauty if it's not, in the end, true? Beauty is truth, and truth is beauty."

I asked Updike if his own attitude toward natural theology was as contemptuous as Barth's was. Some people think there's a God because they have a religious experience. Some think there's a God because they believe the priest. But others want evidence, evidence that will appeal to reason. And those are the people that natural theology, by showing how observations of the world around us might support the conclusion that there is a God, has the power to reach. Is Updike really willing to leave those people out in the cold just because he doesn't like the idea of a God who lets himself be "intellectually trapped"?

Updike paused for a moment or two, then said, "I was once asked to be on a radio program called *This I Believe*. As a fiction writer, I really don't like to formulate what I believe because, like a quantum phenomenon, it varies from day to day, and anyway there's a sort of bad luck attached to expressing yourself too clearly. So . . . ah . . . wait, my wife is just showing me a big thermometer . . . numbers everywhere . . . now what was I saying? Oh yes, on this radio program I conceded that ruling out natural theology does leave too much of humanity and human experience behind. I suppose even a hardened Barthian might cling to at least one piece of natural theology, Christ's saying, 'By their fruits shall ye know them'—that so much of what we construe as virtue and heroism seems to come from faith. But to make faith into an abstract scientific proposition is to please no one, least of all the believers. There's no intellectual exertion in accepting it. Faith is like being in love. As Barth put it, God is reached by the shortest ladder, not by the longest ladder. Barth's constant point was that it is God's movement that bridges the distance, not human effort."

And why should God make that movement? Why should he have created a universe at all? I remembered Updike saying somewhere that God may have brought the world into being out of spiritual fatigue—that reality was a product of "divine acedia." What, I asked him, could this possibly mean?

"Did I say *that*? God created the world ~~out of boredom~~? Well, Aquinas said that God made the world 'in play.' *In play*. In a playful spirit he made the world. That, to me, seems closer to the truth."

He was silent again for a moment, then continued. "Some scientists who are believers, like Freeman Dyson, have actually tackled the ultimate end of the universe. They've tried to describe a universe where entropy is almost total and individual particles are separated by distances that are greater than the dimensions of the present observable universe . . . an unthinkably dreary and pointless vacuum. I admire their scientific imagination, but I just can't make myself go there. And a space like that is the space in which God existed and *nothing else*. Could God then have suffered boredom to the point that he made the universe? That makes reality seem almost a piece of light verse."

What a lovely conceit! Reality is not a "blot on nothingness," as Updike's character Henry Bech had once, in a bilious moment, decided. It is *a piece of light verse*.

I told Updike how much I had enjoyed the chat. He said he had been almost out of breath at the beginning because he had just come in from playing kickball with his grandchildren. "I find when I play kickball, which I did with ease most of my life, that at seventy-five it's a definite strain," he said, laughing. "You listen to your heart beating and hear your own rasping lungs. It's a good way to keep in touch with what stage of life you're at."

A few months later, Updike was diagnosed with lung cancer. Within a year he was dead.

14

THE SELF: DO I REALLY EXIST?

> I am, however, a real thing and really exist. But what thing?
> I have answered: a thing that thinks.
>
> —DESCARTES, Meditations

Why is there Something rather than Nothing? I thought I finally had the answer. It had come in the shape of a proof, almost geometrical in style, one that Spinoza might have found congenial. And Sherlock Holmes might have found my proof congenial too, for it proceeded in precisely the way that, as Holmes insisted to his faithful but less quick-witted companion, Dr. Watson, good detective work ought to proceed: "How often have I said to you that when you have eliminated the impossible, whatever remains, *however improbable*, must be the truth?"

And the last line of my proof not only assured the existence of a nonempty reality. It also dictated the general form that this reality is bound to take: that of infinite mediocrity. If the principles underlying my reasoning were correct, the world must be as far removed from encompassing absolutely everything as it is from encompassing absolutely nothing. But that conclusion raised a new perplexity. If the world falls infinitely short of ontological completeness, then why am *I* a part of it? How did I happen to make the existential cut? And why do I find myself somewhat giddy at the thought that I did?

The brute fact of my existence might not have been so mysterious if this world, among all rival realities, had been marked out for actuality by some special feature. In that case, I might have explained my personal existence by appeal to that special cosmic feature. Suppose, for example, that the cosmos existed because it satisfied an abstract need for goodness, the way John Leslie believed. On that axiarchic/Platonic view, I must be here because my own existence contributes a little jot of goodness to the cosmic total. Or suppose more fancifully that the cosmos is, as John Updike had suggested, like a "bit of light verse." Then the rationale for my own life might be the role that I played in the cosmic metrical scheme, or even in the cosmic joke. Any such special feature, one that marked this particular world out for existence, would lend meaning to my own existence as an element of that world. It would endow my life with a cosmic purpose: to be as ethically good as possible, or to be as lightly poetical as possible, or whatever.

But reality has no such special feature. So, at least, my quest for complete ontic understanding had led me to conclude. The existence of this cosmos can be fully explained only on the assumption that it is middling in every way—a vast Walpurgisnacht of mediocrity. Even its infinity is a middling one, since the infinite cosmos still falls infinitely short of achieving utter plenitude. It is like a randomly chosen subset of the natural numbers, a subset that includes infinitely many numbers but also leaves infinitely many numbers out.

And if reality has no special feature, then my own presence in it cannot be explained by the hypothesis that I somehow enhance that feature, add something to it. Thus there can be no cosmic point to my existence—or rather, the only point to my existence is *that I exist*. Sartre was on to something like this when he said, "Existence precedes essence." And the purpose of my life? As eponymous antihero of Ivan Goncharov's great novel *Oblomov* is wisely told by his friend Stolz, "The purpose is to live." That is a tautology worth remembering.

So my existence, from the perspective of the cosmos, has neither meaning nor purpose nor necessity. (And that is nothing to be ashamed of. For the same would be true of God, if God existed.) I am an accidental, contingent thing. I might easily not have existed at all.

How easily? Let's do a little calculation. As a member of the human species, I have a particular genetic identity. There are about 30,000 active genes in the human genome. Each of these genes has at least two variants, or "alleles." So the number of genetically distinct identities the genome can encode is at least 2 raised to the thirty-thousandth power—which roughly equals the number 1 followed by 10,000 zeros. That's the number of potential people allowed by the structure of our DNA. And how many of those potential people have actually existed? It is estimated that about 40 billion humans have been born since the emergence of our species. Let's round the number up to 100 billion, just to be on the conservative side. This means that the fraction of genetically possible humans who have been born is less than 0.00000 . . . 000001 (insert about 9,979 extra zeroes in the gap). The overwhelming majority of these genetically possible humans are unborn specters. Such is the fantastic lottery that I—and you—had to win in order to shimmer on to the scene. This is contingency with a vengeance.

The fact that we prevailed against these stupefyingly long odds makes us "the lucky ones"—so says Richard Dawkins. Sophocles evidently disagreed. "Never to be born is best of all," the chorus in *Oedipus at Colonus* declares. Bertrand Russell took a more agnostic position on this matter, writing, "There is a general belief (which I have never understood) that it is better to exist than not to exist; on this ground children are exhorted to be grateful to their parents." If your parents had never met, of course, you would not exist. But much more than the mere meeting of your parents, or even their sexual congress at a particular moment in history, had to go improbably right in order for you to see the world. Perhaps the entity that really deserves your gratitude is not your mother or your father, but rather the plucky little sperm that, carrying half of your genetic identity as its cargo, gamely made its way through the amniotic sea, past millions of its ejaculate-rivals, to unite with the egg.

The coming-into-being of my genetic identity was indeed a long shot. But was even *that* enough to ensure the coming-into-being of *me*? Could this genetic identity not just as easily produced not me, but, as it were, my identical twin? (If you happen to be one of a pair of identical twins, try this thought experiment. Imagine that the zygote that split apart shortly after fertilization to produce you and your twin had instead remained a single clump of cells. Would the unique baby born to your parents nine months later have been you? Your twin? Neither?)

And am I really nothing more than an instance of the genetically defined species *Homo sapiens*? I certainly seem to be able to imagine my self migrating into some nonhuman form—a penguin, perhaps, or a robot, or an immaterial being like an angel. So maybe I am not essentially a biological organism after all. Maybe I am essentially something else.

Although I am not sure about what I ultimately am, there is one thing I do know: *I exist*. This proposition may be a contingent truth, but it is also an *a priori* one. I cannot deny it without contradicting myself. (I might deny it jocularly, but that would only be to say that I am economically or socially negligible, not that I am a metaphysical zero.) Even in the extremity of doubt about the world, the fact of my existence is a beacon of certitude. So, at least, Descartes insisted. *Cogito ergo sum*, he put it in his famous phrase: "I am thinking, therefore I exist." And from the claim that his existence was self-evident from the mere fact that he was thinking, Descartes moved directly to the stronger claim that he was, in essence, a *thinking being*—that is, a pure subject of consciousness. As such, the "I" in "I am thinking" had to refer to something that was distinct from his physical body—something immaterial.

Did Descartes here infer more than he was entitled to? As many commentators have pointed out (beginning with Georg Lichtenberg in the eighteenth century), the "I" in his ultimate premise is not quite legitimate. All Descartes could assert with certainty was "There are thoughts." He never proved that thoughts require a thinker. Perhaps the pronoun "I" in his proof was just a misleading artifact of grammar, not a name for a genuinely existing thing.

Suppose you turn your attention inward in search of this *I*. You may encounter nothing more than an ever-changing stream of consciousness, a flow of thoughts and feelings in which there is no real self to be discovered. That, at least, was what David Hume, a century after Descartes, found when he conducted his own introspective experiment. In his *Treatise of Human Nature*, Hume wrote, "When I enter most intimately into what I call *myself*, I always stumble on some particular perception or other, of heat or cold, light or shade, love or hatred, pain or pleasure. I never catch *myself* at any time without a perception, and never can observe anything but the perception.... If anyone, upon serious and unprejudiced reflection, thinks he has a different notion of *himself*, I must confess I can reason no longer with him."

So who is right, Descartes or Hume? Is there an *I* or not? And if there is not, just what am I wondering about when I wonder why I exist?

Even today, the nature of the self is an issue that divides and perplexes philosophers. Perhaps a slight majority lean toward Hume's view that the enduring self is something of a fiction, a shadow cast by the pronoun "I." Derek Parfit, for example, likens the self to a club, one that might change its membership over time, disband altogether, and then reconvene under the same name in a different form. Daniel Dennett says that "selves are not independently existing soul-pearls, but artifacts of the social processes that create us." Galen Strawson thinks that, within each person's stream of consciousness, little transient selves constantly wink in and out of existence, none of them lasting for more than an hour or so. "There simply isn't any 'I' or self that goes on through (let alone beyond) the waking day," Strawson claims, "even though there's obviously an 'I' or self at any given time." Moreover, whatever pop-up self happens to be hanging around at the end of each day is soon, according to Strawson, annihilated by the oblivion of sleep. Each morning awakes a new Cartesian "I."

Even Thomas Nagel, who tends to take a robustly realist view of the self, thinks its true nature could be partly hidden from us. "I may understand and be able to apply the term 'I' to myself without knowing what I really am," Nagel has written.

If the inner *I* is elusive, perhaps there's a reason for that. What, after all, is the self supposed to be? In modern—that is, post-Cartesian—thought, philosophers have laid down two broad conceptual requirements that the self must meet. First, whatever else it is, the self is the *subject of consciousness*. The various experiences I am having at a particular moment—seeing a patch of blue sky through the window, hearing a distant siren, feeling a slight headache, thinking about lunch—are part of the same consciousness because they belong to the same self. I can identify the headache-y feeling as my own without any chance of being mistaken. (Hence the absurdity, in Charles Dickens's *Hard Times*, of the sick-bed statement made by Mrs. Gradgrind: "I think there is a pain somewhere in the room, but I couldn't positively say that I have got it.") And the second requirement is that the self be capable of *self-consciousness*—of being aware of itself, of having "me-ish" experiences.

But isn't there a fatal tension between these two requirements? How can one and the same thing be simultaneously both the *subject* and the *object* of consciousness? The very idea struck Schopenhauer as "the most monstrous contradiction ever thought of." Wittgenstein agreed. "The I is not an object," he declared. "I objectively confront every object. But not the I." Like Schopenhauer before him, Wittgenstein compared the *I* to the eye. Just as the *I* is the source of consciousness, the eye is the source of the visual field. But the eye is not *in* the visual field. It cannot see itself.

That may be why Hume was unable to find his own self. It may also be why (as Nagel thought) I can't really know what I am.

Still, I do seem to be asserting something when I say, "I exist." And the content of my assertion must be different from the content of yours when you utter the same sentence. But how? What makes one subject of consciousness different from another?

One view is that the contents of consciousness are what constitute the self. This is the *psychological* criterion of self-identity. On this view, to say "I exist" is just to assert the existence of a certain more or less continuous bundle of memories, perceptions, thoughts, and intentions. What makes me *me* and you *you* is our distinctive bundles.

But what happens if I undergo amnesia and lose all my memories? Or what if a fiendish neurosurgeon manages to erase all my memories and replace them with your memories? And what if he performed the reverse operation on you? Would we find ourselves waking up in each other's body?

If you think that the answer to the last question is "yes," consider the following scenario. You are informed that you are going to be tortured tomorrow. Understandably, this makes you fearful. But prior to the torture, you are told, your memories will be wiped out by the fiendish neurosurgeon and replaced with my memories. Would you still have reason to fear the torture? If you did, it would mean that, despite your complete psychological makeover as me, it would still be *you* who endured the pain.

Such a thought experiment was proposed in 1970 by the philosopher Bernard Williams to show that the psychological criterion of personal identity must be mistaken. But if psychological factors don't determine my self-identity, what could? The obvious alternative—endorsed by Williams and later, more tentatively, by Thomas Nagel—is the *physical* criterion. My identity as a self is determined by my body; or, more specifically, by my brain, the physical object that is causally responsible for the existence and continuity of my consciousness. On the "I am my brain" view, the actual contents of your stream of consciousness don't matter to your identity. What is all-important is the particular blob of gray meat that is lodged in your skull. You cannot survive the destruction of this blob. Your self could not be "uploaded" into a computer, nor could it be resurrected in some ethereal form. Nagel has gone so far as to suggest that even if an *exact physical replica* of your brain were created, and then stocked with your memories and lodged in a clone of your body, the result would still not be *you*. (Although it would certainly *think* it was you.)

So when I say "I exist," I may just be asserting the existence of a particular (functioning!) brain. Then the question *Why do I exist?* has a purely physical answer: I exist because, at a certain moment in the history of the universe, a certain bunch of atoms happened to come together in a certain way.

The problem with this easy answer, as Derek Parfit has pointed out, is that even the physical identity of my brain is not an all-or-nothing affair. Suppose, Parfit says, all of your brain cells have some defect that would eventually be fatal. Suppose further that a surgeon could replace these brain cells with duplicates that were not defective. The surgeon might do this gradually, in a series of, say, a hundred cell-grafting operations. After the first operation, 99 percent of your original brain would remain. In the middle of the series, half of your brain would consist of original cells and half of duplicates. And just before the final operation, your brain would be a 99 percent replica. Would the self at the end of this series of operations still be you, even though your original brain had been completely destroyed and replaced? And if it ceased to be you, at what point in the series did you suddenly disappear and get replaced by a new self?

It looks as if neither the psychological criterion nor the physical criterion decisively settles who I am. This raises a disturbing suspicion. Perhaps there is *no fact of the matter* when it comes to my identity. Perhaps there is no real answer to the question of whether I exist or not. Even though I am referring to something when I say "I" or "me," that something has no ontic solidity. It does not figure among the True and Ultimate Furniture of the Universe. It has no existence apart from the constantly shifting mental states that populate my mind and the constantly changing set of physical particles that constitute my body. The self is, to use Hume's analogy, like a nation; or, to use Parfit's analogy, like a club. We can keep track of its identity from moment to moment. But the question of whether it remains the same over long stretches of time, or through great physical and psychological discontinuities, is an indeterminate, even an empty one. The enduring, substantial, self-identical *I* is a fiction. As the Buddha put it, the self is "only a conventional name given to a set of elements."

Hume found this conclusion, convinced though he was of its truth, to be a depressing one. It left him, he wrote, "in the most deplorable condition imaginable, inviron'd with the deepest darkness." (Fortunately, he was able to find relief by playing backgammon with friends.) By contrast, Derek Parfit, rather more like the Buddha, finds it "liberating and consoling." Before, when Parfit thought that the existence of his self was a deep all-or-nothing fact, "My life seemed like a glass tunnel, through which I was moving faster every year, and at the end of which there was darkness." Once liberated from the self, however, "the walls of my glass tunnel disappeared. I now live in the open air."

Suppose that the Cartesian *I*—the *I* of "I think, therefore I am"—truly is an illusion. How could such an illusion arise? (And, one might want to ask, for *whom* or *what* is it an illusion?) Well, to be an *I* is to have self-consciousness, to enjoy the power of reflexive awareness. So perhaps the *I* is conjured into existence in the very act of thinking about itself. Perhaps, in other words, the *I* is self-creating!

Such was the daring hypothesis that Robert Nozick put forward, albeit "with great hesitation," to deal with the otherwise "quite intractable" problem of the source of the self. According to Nozick, when the Cartesian says, "I think," he is not referring to any preexisting entity. Nor is he truly describing an already existing state of affairs. Rather, the state of affairs is *made true* by the declaration. The entity referred to by the pronoun "I" is (somehow) delineated in the very act of self-reference, which picks out "the thing of greatest organic unity" that includes the act itself. And what are the boundaries of this organically unified self-creation? "Nothing we have said thus far limits what the self-synthesizing self can synthesize itself as," Nozick observed. He even entertains the possibility that this self might be "identical with the underlying substance of the universe, as in the Vedanta theories that Atman is Brahman."

Once you entertain the notion that the *I* is self-creating, it's easy to find yourself sliding down a slippery transcendental slope. And what lies at the bottom of that slope is a curious form of idealism, one which says that the *I*, in creating itself, creates *all of reality*. Daft as this notion might sound, it crops up repeatedly in European philosophy after the time of Kant. Versions of it can be found in Hegel, Fichte, and Schelling in the nineteenth century, and in Husserl and Sartre in the twentieth century.

Take Johann Gottlieb Fichte, the son of an impoverished ribbon-weaver who grew up to become not only the esteemed philosophical successor to Kant but also the intellectual father of German nationalism. Fichte held (like Nozick) that the *I* leaps into existence in the very act of "positing" itself. The statement "I = I," being an instance of the logical law of identity, is a necessary truth. Indeed, according to Fichte, it is the *only* necessary truth, because it presupposes nothing. (Ordinarily, the truth of the identity "A = A" presupposes existence of A. But the existence of the *I* in "I = I" is guaranteed by the self-positing nature of the self.) As the sole necessary truth, "I = I" must be the ground of all other knowledge. Therefore, Fichte thought, all knowledge must ultimately be *self-knowledge*. The transcendental subject not only creates itself in the act of positing, but also creates the world—a true ontological tour de force! "All art, religion, science and institutions are gathered into this process, expressing some part of the great spiritual journey, whereby the empty I = I takes on flesh, so as to know itself at last as an ordered and objective reality, and also free," as the contemporary philosopher Roger Scruton described Fichte's miraculous world-engendering dialectic.

Edmund Husserl, founder of the phenomenological movement in the early twentieth century, endowed the *I* with similar ontic powers. "The objective world," Husserl submitted, "derives its whole sense, and its existential status . . . from me myself, *from me as the transcendental Ego.*"

Now, for me to believe that I am literally the fountainhead of all reality would be metaphysical hubris, if not insanity. Yet whatever my self really is—a substance, a bundle, a locus, a receptacle, a vehicle, a self-writing poem, a grammatical shadow, a transcendental Ego—it does seem to stand at the center of the world. "The world is *my* world," Wittgenstein declared in proposition 5.62 of the *Tractatus.* And he strengthened the point in proposition 5.63: "I am my world. (The microcosm.)"

Of course, the world could be *my* world—as opposed to *your* world, or *her* world—only if I am the sole genuine self: the Metaphysical Self. Not being a solipsist, I don't believe this. (Although, as a child, I believed I could make the world go dark by closing my eyes.) Even though I am at the center of my subjective world, I believe there is an objective world, one which exists quite independently of me, a vast expanse of space and time that I know a relatively tiny portion of. This objective world was there before I was born, and it will continue to be there after I die. I also believe that the objective world is centerless. It has no built-in perspective—the way it would, say, if it existed in the mind of God. And it is as centerless that I must try to understand the world.

Thomas Nagel has memorably dubbed this centerless view of reality "the view from nowhere." And the self that strives to take this view of reality he calls the "objective" or the "true" self. The objective self, on Nagel's account, is something different from a particular person. This self treats the experiences of that particular person as a sort of window onto the world, using those experiences to construct a perspectiveless conception of reality. But, having done this, the objective self is confronted with a startling puzzlement: "How can I, who am thinking about the entire, centerless universe, be anything so specific as *this*: this measly, gratuitous creature existing in a tiny morsel of spacetime, with a definite and by no means universal mental and physical organization? How can I be anything so *small* and *concrete* and *specific*?"

When he considers the world objectively, Nagel is astonished that his consciousness should be localized in a particular human being. "What kind of a fact is it," he asks, "that I am Thomas Nagel?" It seems to him miraculous that he, this fleeting organic bubble in the ocean of reality, should be "the world soul in humble disguise." Lest this sound like metaphysical megalomania, Nagel pleads in extenuation that "the same thought is available to any of you. You are all subjects of the centerless universe, and mere human or Martian identity should seem to you arbitrary. I am not saying that I individually am the subject of the universe: just that I am *a* subject that can have a conception of the centerless universe in which TN is an insignificant speck, who might easily never have existed at all."

Philosophers who would like to deflate Nagel's "objective self" argue that the sentence "I am TN" is true if and only if it is uttered by TN, and that there is nothing more to it, startling or otherwise, than that. It's no different from the sentence "Today is Tuesday," which is true if and only if it's uttered on Tuesday. But Nagel counters that such an impersonal semantic analysis leaves a gap in our conception of the world. Even when all the public information about the human being TN has been included in the objective conception, he says, "the additional thought that TN is *me* seems clearly to have further content. And it is important that the content is startling."

(At lunchtime, after typing the preceding words, I went around the corner to the local gourmet market in Greenwich Village to get a chicken-and-avocado sandwich. And there, standing unobtrusively in the checkout line with a basket of groceries, was Thomas Nagel himself—the world soul in humble disguise. I nodded to him, and he nodded back amiably.)

Do I feel similarly startled that *I* am JH? It depends on my mood. Sometimes the thought strikes me as deeply mysterious. Sometimes it strikes me as perfectly vacuous. (In that respect, it's a little like the thought that there is Something rather than Nothing.) Unlike Nagel, I don't feel particularly surprised when I ponder my cosmic unimportance. I don't have any difficulty seeing myself as an "insignificant speck."

Could I have been someone else other than JH—some quite different speck? Suppose the history of the world had gone just as it actually has, only I was Napoléon and Napoléon was me. What happens when I try to conceive of such a possibility? Well, I might imagine myself as being short of stature and wearing a cockaded hat, my hand thrust into my tunic, viewing the devastated battlefield of Austerlitz. But, as the philosopher Bernard Williams shrewdly pointed out, all I have really imagined here is play-acting at being Napoléon. This no more enables me to understand how I could have *been* Napoléon than seeing Charles Boyer play Napoléon on the screen enables me to understand how *he* could have been Napoléon.

If I say to myself, "I might have been Napoléon," the pronoun "I" cannot very well refer to the empirical JH, the one who led a quiet and harmless existence on the North American continent in the late twentieth and early twenty-first centuries. For in that case, the proposition would be a straight contradiction. So the "I" here must refer to my self as it would be if it were stripped of all its physical and psychological baggage—to my pure, timeless, featureless Cartesian ego. *That* is the self that I am trying to imagine switching with Napoléon. But do I have such a self? Do you?

If you do, then that opens up possibilities even more vertiginous than switching selves with Napoléon. For instance, you might (as Derek Parfit has observed) suddenly cease to exist in the middle of reading this paragraph, and be replaced by a new self that would inhabit your body and assume your precise psychological makeup. If such a thing happened, there would never be any external evidence of it.

Another possibility is that the world could be exactly as it is, except that your pure Cartesian ego never existed at all. Your empirical person, with its genetic identity, memories, social relations, and all the rest of its life history would still be there. Only it would not be *you*. It would be your (perfectly) identical twin. The little light of your consciousness would never have flickered in the world.

It is hard to find a philosopher these days who takes seriously the idea of such a pure Cartesian self. Parfit deems it "unintelligible," and Nagel, despite his talk of the "objective self," gives no indication that he thinks such a self can be wholly detached from its physical and psychological moorings. (Indeed, if the brain is the core of the self, as Nagel tentatively believes, then even transplanting my brain into Napoléon's body would leave me stuck with being just a shorter and sallower JH.) And if the self could be so detached, Williams asked, what would be left to distinguish one pure Cartesian ego from another? What exactly would be subtracted from the world by the removal of *me*?

"AMAZEMENT THAT THE universe should have come to contain a being with the unique property of being me is a very primitive feeling," Nagel observes. Like him, I cannot help feeling somewhat astonished that I exist—that the universe has come to produce these very thoughts now bubbling up in my stream of consciousness.

Yet the astonishment I feel at my improbable existence has a curious counterpoint: the difficulty I have in imagining my sheer *nonexistence*. Why is it so hard to conceive of a world without me, a world in which I never put in an appearance? I know, after all, that I am hardly a necessary feature of reality. Still, like Wittgenstein, I can't think about the world without thinking of it as *my* world. Although I am part of reality, reality feels like a part of me. I am its hub, its epicenter, the sun that illumines it, the baby to its bathwater. To imagine that I never existed would be like imagining that the *world* never existed—that there was Nothing rather than Something.

The feeling that the "something-ness" of reality depends on my existence is, I know, a solipsistic illusion. Yet even when it is recognized as such, it retains a considerable grip. How can I loosen its grip? Perhaps by holding steadily before me the thought that the world got on quite happily for many eons prior to that unlikely moment when I was abruptly awakened to life out of the night of unconsciousness, and that it will get on quite happily after the inevitable moment to come when I return to that night.

15

RETURN TO NOTHINGNESS

A man finds himself, to his great astonishment, suddenly existing,
after thousands of years of non-existence; he lives for a little while;
and then, again, comes an equally long period when he must exist
no more. The heart rebels against this, and feels that
it cannot be true.

—ARTHUR SCHOPENHAUER, *"The Vanity of Existence"*

lthough my birth was contingent, my death is necessary. Of that I am reasonably sure. Yet I find my death difficult to imagine. And here I am in impressive company. Freud said he could not conceive of his own death. So did Goethe before him. "It is entirely impossible for a thinking being to think of its own non-existence, of the termination of its thinking and life," Goethe said, adding that "to this extent, everybody carries within himself, and quite involuntarily at that, the proof of his own immortality."

Such a "proof" of immortality is, alas, quite worthless. It is another instance of what has been called the *philosopher's fallacy*: mistaking a failure of imagination for an insight into reality. Not everyone, moreover, finds his own death inconceivable. Lucretius argued, in the stately verse of *De Rerum Natura*, that it is no harder to imagine not existing after one's death than it is to imagine not existing prior to one's birth. David Hume evidently felt the same way. Indeed, he claimed to find posthumous nonexistence no more frightening than prenatal nonexistence. When asked by James

Boswell whether the thought of annihilation terrified him, Hume calmly replied, "Not the least."

To exhibit such sangfroid in the face of death is said to be "philosophical." To philosophize, Cicero declared, is to learn how to die. Here Socrates is held to be the model. Sentenced to die by an Athenian court on the charge of impiety, Socrates serenely and willingly drank the fatal cup of hemlock. Death, he told his friends, might be annihilation, in which case it is like a long, dreamless slumber; or it might be a migration of the soul from one place to another. Either way, it is nothing to be feared.

Why should the prospect of annihilation disturb me, if it did not disturb Socrates or Hume? I have already said that I cannot easily imagine my own death. That might make death seem a mysterious and therefore fearful thing. But I cannot imagine my being totally unconscious either, yet I enter that state every night, and quite fearlessly.

It is not the prospect of unending nothingness as such that makes death terrifying; it is the prospect of losing all the goods of life, and losing them permanently. "If we are to make sense of the view that to die is bad," Thomas Nagel has written, "it must be on the ground that life is a good and death is the corresponding deprivation or loss." And just because you don't experience the loss once you've ceased to exist doesn't mean the loss is not bad for you. Suppose, Nagel says, an intelligent person has a brain injury that reduces him to the mental condition of a contented baby. Certainly this would be a grave misfortune for the person, even though it would not be experienced as such. Then is not the same true for death, where the loss is still more severe?

But what if your life contains no goods? What if it's a life of unremitting agony or unendurable tedium? Isn't nonexistence then preferable?

I tend to have conflicting intuitions on this question. But I am impressed by the reasoning of the late British philosopher Richard Wollheim, who maintained that death is a misfortune even when life is utterly devoid of enjoyments. "It is not that death deprives us of some particular pleasure, or even of pleasure," Wollheim wrote. "What it deprives us of is something more fundamental than pleasure: it deprives us of that thing which we gain access to when, as persisting creatures, we enter into our present mental states.... It deprives us of phenomenology, and, having once tasted phenomenology, we develop a longing for it which we cannot give up: not even when the desire for cessation of pain, for extinction, grows stronger."

And I am even more impressed with the testimony that Miguel de Unamuno gave in the *Tragic Sense of Life*:

> I must confess, painful as the confession may be, that even in the days of my youth's simple faith, I never was made to tremble by descriptions of hellfire, no matter how terrible, for I felt, always, that the idea of nothingness was much more terrifying than Hell. Whoever suffers lives, and whoever lives in suffering still loves and hopes, even though over the portal of his abode is written "Abandon all Hope!" And it is better to live in pain than peacefully cease to be at all. The truth is that I could not believe in this atrocious Hell, an eternity of punishment, nor could I imagine a more authentic Hell than that of nothingness and the prospect of it.

THE DREAD OF death goes beyond the thought that the rush of life will continue without us. For even the solipsist, who thinks the world depends on him for its existence, fears death. Nor would my own fear of death be lessened if I thought that I was going to die as the result of some general disaster that wiped out all life on Earth, or even one that obliterated the entire cosmos. In fact, that would make me dread my death all the more keenly.

No, it is the prospect of *nothingness* that induces in me a certain queasiness—if not, as it did with Unamuno, outright terror. How to envisage this nothingness? From the objective standpoint, my death, like my birth, is an unremarkable biological event, one that has happened billions of times to members of my species. But from the inside it is unfathomable—the vanishing of my conscious world and all that it contains, the end of subjective time. This is my "ownmost death," as the American philosopher Mark Johnston terms it, the snuffing out of my very self, the "end of this arena of presence and action." The prospect of one's ownmost death is perplexing and terrifying, Johnston submits, because it reveals that we are not, as we supposed, the fountainhead of the reality we inhabit, the center of the world.

Nagel makes a similar point. From the inside, he writes, "my existence seems to be a universe of possibilities that stands by itself, and therefore stands in need of nothing else to continue. It comes as a rude shock, then, when this partly buried self-conception collides with the plain fact that TN will die and I with him. This is a very strong form of nothingness. . . . It turns out that I am not the sort of thing I was unconsciously tempted to think I was: a set of ungrounded possibilities as opposed to a set of possibilities grounded in a contingent actuality."

Not all philosophers regard the inexorable return to nothingness in such a depressing light. Take Derek Parfit, whose theorizing about the insubstantial nature of the self liberated him from the belief that his own continued existence was an all-or-nothing affair. His death, he believes, will merely break some psychological and physical continuities, while leaving others intact. "That is all there is to the fact that there will be no one living who will be me," Parfit writes. "Now that I have seen this, my death seems to me less bad."

Less bad—well, that's some progress. But is there nothing *positive* that can be said for nothingness? What about the ideal of Nirvana, the blowing out of the flame of the self, the cessation of desire? Could the personal extinction that death holds out to us be a state of perpetual peace, as Buddhist philosophy maintains? But how can you enjoy something if you do not exist? Hence the wag's definition of Nirvana: having just enough life to enjoy being dead.

Schopenhauer, influenced by Buddhist thought, proclaimed that all will is suffering. Therefore, the ultimate goal of the self should be annihilation— a return to the unconscious eternity whence it emerged: "Awakened to life out of the night of unconsciousness, the will finds itself as an individual in an endless and boundless world, among innumerable individuals, all striving, suffering, and erring; and, as if through a troubled dream, it hurries back to the old unconsciousness."

Schopenhauer's quasi-Buddhist view of life may seem a needlessly jaundiced one. Still, the idea of annihilation as a return to a lost state of peace can have a powerful emotional resonance, one that harkens back to our childhood. We come into existence in the womb—a warm sea of unconsciousness—and then find ourselves at our mother's breast, in a consummate state of satisfied desire. As our sense of self gradually emerges, it is in an atmosphere of total dependence on our parents—a dependence that is more prolonged in the human species than in any other. As adolescents, we must shed this dependence by rebelling against our parents, repudiating the comforts of home, and striking out into the world. There we compete to reproduce ourselves, beginning the cycle anew. But the world is a dangerous place, full of strangers; and our rebellion against our parents leaves us with a sense of alienation, a sense of having ruptured a primal bond. Only by returning home can we expiate our crime of existence, achieve reconciliation, and restore oneness.

What I have just presented is a caricature of Hegel's dialectic of the family. Crude as it is, it does give some psychological sense to the feeling that reality—the world outside the womb of the family, the world of becoming—is a place of alienation. "We are not at home in the world, and thus homelessness is a deep truth about our condition," writes Roger Scruton, commenting on the idea of existential alienation. "Here, indeed, is the root of original sin: through consciousness, we 'fall' into a world where we are strangers." Hence our deep-seated desire to return to "the primordial point of rest": the landscape of childhood and the safety of the family hearth.

And what is the endpoint of this longed-for journey of expiation, atonement, and restored unity? That warm maternal sea from which we emerged—that eternal home of contented unconsciousness. Nothingness.

It was while I was in the midst of entertaining such seductively woolly notions that I got some news. My mother was about to die.

THIS NEWS CAME somewhat abruptly, but it was not entirely unexpected. A month and a half earlier, my mother, who lived in the Shenandoah Valley in Virginia, where I myself had been born, went to the hospital with what appeared to be a nastily persistent case of bronchitis. A tumor was found on her lung. Up to that point, she had enjoyed robust good health throughout the seven-plus decades of her life, even winning a local tennis tournament a few years ago. But with the diagnosis of cancer, her condition deteriorated with awful swiftness. Within a week, her legs began to grow numb and paralyzed. The tumor, it turned out, had metastasized to her spinal column. Daily radiation treatments proved useless. There was nothing else the doctors could do. So my mother was transferred to a hospice.

She was very happy in the hospice for the first couple of weeks. It was a small, homey place, lying in a secluded meadow, with a view of the Blue Ridge Mountains. The people who took care of her were nice, she said, and the food was good—plenty of bacon for breakfast. My mother phoned me every day in New York. Dear friends were visiting her. She was following the French Open tennis tournament on TV. She wasn't in much pain. (How much morphine were they giving her?) And she didn't seem at all afraid of death. She had been a devout Catholic all her life, attending daily mass and saying the Rosary every morning, among other devotions. She had led a good life and kept all the commandments, so she was sure she would be going to heaven. There she would see my father, who had died quite suddenly in his sleep of a heart attack a decade earlier, after a vigorous day of tennis and swimming in the sea, and probably also my younger brother, who had died a few years ago at a party after taking too much cocaine.

I thought my mother might be around for a while—the doctors had given her six months. But then, early one morning, a nurse called. My mother had suddenly taken a bad turn. She had stopped eating. She was unable to drink fluids, which she simply choked on. (She had given instructions that she did not wish to be hydrated intravenously.) There was a rattle in her throat when she slept. And she rarely awoke. It looked as if she would die within a few days.

So I immediately borrowed a car and made the eight-hour drive down from New York to Virginia. When I got to the hospice that evening, a priest was in my mother's room, a young, grinning Filipino who spoke bad English but seemed holy in his way. He had performed last rites for my mother and given her absolution for her sins. When I stood over her bed, her eyes opened, and she seemed to recognize me. Trying to think of something lighthearted to say, I told the priest that my mother had now received every sacrament except holy orders, which put her one sacrament ahead of him. My mother's eyelids fluttered, and she smiled.

The next day I spent sitting by her bedside, holding her hand, saying, "It's Jim, I'm here with you, I love you," over and over again. She drifted in and out of consciousness. At one point some people from her church came into the room and started chanting over her bed an annoyingly repetitive prayer to the Blessed Virgin Mary. When they finally went away, I noticed that my mother's mouth looked very dry. I put some cool water on a swab and dabbed it on her lips. Her eyelids fluttered open and she looked at me. "You have a handsome forehead," she said, in a whispery voice. ("Thank you!" I replied.) Then her eyes closed again. After a few hours I left, doubting that she'd make it through the night.

But when I got back early the following morning my mother was still alive. Her eyes were closed. She had not regained consciousness during the night, the nurse told me. She no longer reacted to the sound of my voice. I was alone with her. I put my hand on her brow. I gave her a kiss on the cheek. She was breathing steadily, and her facial muscles looked relaxed—no sign of pain. I sang a corny song called "True Love" that she and my father used to sing to each other in harmony, amid gales of laughter. I talked about trips we had taken together as a family many years ago. Not the slightest response. I looked out the french doors of her room at the summer flowers outside, the birds, the butterflies. Such a sweet scene. Around noon the nurse came in to shift my mother's body in bed. Her legs were mottled, showing the circulation had stopped, and the mottling was advancing up her body. "She has maybe an hour to live," the nurse told me, and left the room.

My mother's breathing was getting shallower. Her eyes remained closed. She still looked peaceful, although every once in a while she made a little gasping noise.

Then, as I was standing directly over her, still holding her hand, my mother's eyes opened wide, as if in alarm. It was the first time I had seen them that day. She seemed to be looking at me. She opened her mouth. I saw her tongue twitch two or three times. Was she trying to say something? Within a couple of seconds, her breathing stopped.

I leaned down and whispered that I loved her. Then I went into the hall and said to the nurse, "I think she just died."

I returned to the room to be alone with my mother's body. Her eyes were still a little open, and her head was cocked to the right. I thought about what was going on in her brain, now that her heart had stopped and the blood had ceased to flow. Deprived of oxygen, the brain cells were frantically but vainly attempting to preserve their functioning until, with gathering speed, they chemically unraveled. Perhaps there had been a few seconds of guttering consciousness in my mother's cortex before she vanished forever. I had just seen the infinitesimal transition from being to nothingness. The room had contained two selves; now it contained one.

A half hour passed before the undertaker, a well-groomed young man in an unseasonable black wool suit, arrived. I gave him instructions and left my mother for the last time.

That night, I treated myself to dinner at a stylish and ambitious new restaurant that a young chef from Manhattan had recently opened in my hometown. I hadn't eaten all day. I drank champagne at the bar and announced to the bartender, rather glibly, that my mother had just died that afternoon. At the table I ordered monkfish and heritage pork and heirloom beets, and I drank a delicious bottle of a locally produced Cabernet Franc. I got a little drunk and bandied jokes with my waitress, who had a genial red face and husky Southern accent. I had something for dessert and a sweet wine to go with it. Then I left the restaurant and walked the deserted downtown streets for a while, admiring the well-preserved mix of pre–Civil War and Victorian architecture that, as a boy, I had taken for granted. My hometown, like Rome, was built on seven hills. I walked to the top of the highest of them and took in the twinkling lights of the surrounding Shenandoah Valley. Then I broke into convulsive sobs.

When I woke the next morning in what had been my mother's house— now weirdly empty, despite the profusion of old furniture and antiques and other debris that she had hoarded—the air outside was of an unusual sweetness. There had been heavy showers overnight, but now they had moved east, well out of the valley. I decided to go out for a run: a run with a purpose. I would reenact the Hegelian dialectic of the family, except I would do it backward. Like the title character in John Cheever's short story "The Swimmer," I would return home. But whereas Cheever's character made the journey homeward by breaststroking his way through an almost contiguous series of suburban swimming pools, I would do it by running past the landmarks of my early life, in reverse chronological order, until I ultimately arrived at the site of my conception. I would be The Jogger.

It was a goofy conceit, but one is hardly at one's subtlest in the immediate aftermath of a parent's death. And what made it goofier was that I could not get the Rolling Stones' song "This Will Be the Last Time" out of my head.

As I headed out, the morning fog was beginning to lift. Before long, I could see the distant Blue Ridge Mountains, sharply etched and quite literally blue in the dawn light. I jogged past my old high school, where I had read Sartre and Heidegger in the library and taken up godless existentialism against the orthodox religion my parents thought they had permanently instilled in me, and where my bad companions had taught me to smoke. I jogged past the sprawling faux-Georgian house with the tennis court out back where we had lived during my adolescent years, and where, in a basement bedroom, my sexual awakening had clumsily commenced one night when my parents were out of town. I jogged past the Catholic church where I had received my first communion and where I had piously confessed my absurd childhood sins, and past the old schoolhouse where the nuns had taught me to emulate Saint Francis, the patron saint of the parish.

By and by, I arrived at the foot of the hill on which, just over the top, stood the little white-brick bungalow where my mother and father had first set up housekeeping after they married. The hill was steeper than I remembered. I had to summon greater and greater effort as I made my ascent—just like, I thought to myself, a particle accelerator has to attain higher and higher energies to re-create the very earliest state of the universe. Finally, I reached the top. There was the old house. I looked in the

window of what had been my parents' bedroom—the scene of the Big Bang
(I forgave myself the execrable pun) that had produced me, or had produced, rather, the little symmetrical blob of protoplasm which, through
a long and contingent series of symmetry-breaking events, issued in the
messy reality that I was today. Ontogeny recapitulates cosmogony. Here
was the ultimate home of my inchoate self. I felt moved, but only for a
moment. My journey back was a cliché, a joke. The house had other occu-
pants. Life had moved on. I would not be reunified with my parents until I,
too, entered the nothingness that had already absorbed both of them. That
was the real eternal home. And now I had a clear run to the Void.

Epilogue

OVER THE SEINE

Paris, shortly before the turn of the millennium. I am invited, through the good graces of a mutual friend, to attend a small party at the Collège de France in celebration of the ninetieth birthday of Claude Lévi-Strauss.

On the appointed evening, I make my way from the sixteenth-century apartment house where I am staying, between Place Maubert and the Seine, and head up the Rue Saint-Jacques toward the Panthéon. I enter the courtyard of the Collège de France, pass by the statue of the now-forgotten Renaissance scholar Guillaume Budé, and go inside. After the stateliness of the courtyard, the interior rooms seem meanly proportioned and a bit shabby. There are a dozen or so distinguished academics at the party, plus a sprinkling of journalists, but no cameras or microphones. Fortified by a couple of glasses of the Burgundy that's being served, I obtain an introduction to Lévi-Strauss himself, who rises with difficulty from his chair and shakes my hand tremulously. The conversation is awkward, owing both to my poor French and to my stunned amazement that I am actually having a vis-à-vis with the greatest French intellectual alive.

A few minutes later, Lévi-Strauss is asked to give a little speech to the party. He talks extemporaneously, without notes, in a slow, stately voice.

"Montaigne," he begins, "said that aging diminishes us each day in a way that, when death finally arrives, it takes away only a quarter or half the man. But Montaigne only lived to be fifty-nine, so he could have no idea of the extreme old age I find myself in today"—which, he adds, was one of the "most curious surprises of my existence." He says he feels like a "shattered hologram" that has lost its unity but that still retains an image of the whole self.

This is not the speech we were expecting. It is intimate, it is about death.

Lévi-Strauss goes on to talk about the "dialogue" between the eroded self he has become—*le moi réel*—and the ideal self that coexists with it—*le moi métonymique*. The latter, planning ambitious new intellectual projects, says to the former, "You must continue." But the former replies, "That's your business—only you can see things whole." Lévi-Strauss then thanks those of us assembled for helping him silence this futile dialogue and allowing his two selves to "coincide" again for a moment—"although," he adds, "I am well aware that *le moi réel* will continue to sink toward its ultimate dissolution."

AFTER THE PARTY I leave the Collège de France and go out into the drizzly Paris night. I walk down the rue des Écoles to the Brasserie Balzar, where I have a nice plate of choucroute and drink the better part of a bottle of Saint-Émilion. Then I head back to my apartment and turn on the TV.

There is a book-chat show in progress, hosted by the familiar French television figure Bernard Pivot. His guests tonight are a Dominican priest, a theoretical physicist, and a Buddhist monk. And they are all grappling with a deep metaphysical question, one that was originally posed three centuries ago by Leibniz: *Pourquoi y-a-t-il quelque chose plutôt que rien?* Why is there Something rather than Nothing?

Each of the guests has a different way of answering this question. The priest, a handsome but unsmiling young man wearing severe wire-frame spectacles and attired in a hooded Dominican habit of pure white, argues that reality had to have had a divine origin. Just as each of us came into existence through an act of our parents, the priest says, so the universe must have come into existence through an act of a creator. *Au fond de la question est une cause première—Dieu.* He adds that God was not the first cause in a temporal sense, since God created time itself. God was behind the Big Bang, but not prior to it.

The physicist is an older fellow with a thick head of white hair, wearing a light-blue sport coat and an improbable Western-style string tie. He is grumpily impatient with all this supernatural nonsense. The existence of the universe is purely a matter of chance quantum fluctuations, he says. Just as a particle and its antiparticle can spontaneously arise out of a vacuum, so too can the seed for an entire universe. So quantum theory accounts for why there is something rather than nothing. *Nôtre univers est venu par hasard d'une fluctuation quantique du vide.* Our universe arose by chance from a quantum fluctuation in the void. And that's the end of it.

The monk, attired in crimson and saffron robes, with bare shoulders and a freshly shaved head, has the most interesting line on the question. He also has the most pleasant demeanor. In contrast to the prim-mouthed young priest and the irritable old physicist, the monk beams happiness. A smile continuously plays about his lips. As a Buddhist, he says, he believes that the universe had no beginning. *Il n'y a pas de début.* Nothingness—*le néant*—could never give way to being, he says, because it is defined in opposition to that which exists. A billion causes could not make a universe come into existence out of what does not exist. That is why, the monk says, the Buddhist doctrine of a beginning-less universe makes the most metaphysical sense. *C'est encore plus simple.*

Vous trouvez? interjects Bernard Pivot, eyebrow arched.

The Buddhist monk genially protests that he is not evading the question of origins. Rather, he is using it to explore the nature of reality. What is the universe, after all? *Ce n'est pas bien sûr le néant.* It is not nothingness. Yet it is something very close: an emptiness—*une vacuité.* Things don't really have the solidity we attribute to them. The world is like a dream, an illusion. But in our thinking, we transform its fluidity into something fixed and solid-seeming. This engenders *le désir, l'orgueil, la jalousie.* Buddhism, by correcting our metaphysical error, thus has a therapeutic purpose. It offers *un chemin vers l'éveil*—a path to enlightenment. And it also resolves the mystery of being. When Leibniz asked, *Pourquoi quelque chose plutôt que rien?* his question presupposed that something really and truly exists. And that's an illusion.

Ah oui? says Pivot, again skeptically arching an eyebrow.

Oui! replies the monk, smiling radiantly.

I SWITCH OFF the TV and go out into the chill Paris night for a stroll and a smoke. Leaving my building, I turn toward the Seine, a short block away. Just across the water looms the back of Nôtre Dame, with its flying buttresses. I walk along the quay downriver for a bit, until I get to the Pont des Arts—my favorite bridge, since it has no traffic and is thus (aside from the buskers) quiet. I go halfway across the bridge, where I pause to light a cigarette and take in the view of Paris at midnight.

Before me stretches a gorgeously illumined patch of the great *vacuité* that the Buddhist monk had spoken of. Is it really an insubstantial dream, an empty illusion? Is it gross and viscous and absurd, as Sartre held, or is it a divine gift, as that Dominican priest was just saying? Or could the whole thing just be an inexplicable quantum fluke?

This *pourquoi quelque chose plutôt que rien* business, I think to myself, it's really awfully mysterious. Worth looking into further. Maybe I should write a book about it some day.

I flick my cigarette butt into the dark waters flowing below and head home.

Philosophy, *n.* A route of many roads leading from nowhere to nothing.

—AMBROSE BIERCE, *The Devil's Dictionary*

Acknowledgments

I am indebted to Adolph Grünbaum, Richard Swinburne, David Deutsch, Andre Linde, Alex Vilenkin, Steven Weinberg, Roger Penrose, John Leslie, Derek Parfit, and the late John Updike, all of whom were kind enough to share their time and thoughts with me. Among those with whom I did not speak directly, it should be obvious that I owe the most to Thomas Nagel, a philosopher I have always revered for his originality, depth, and integrity.

I am also grateful to Samuel Scheffler, whose 2010 seminar on the metaphysics of death I was privileged to attend; to my philosophical confidants Anthony Gottlieb, Ned Block, Paul Boghossian, and Jonathan Adler; to my witty and industrious intern, Jimmy O'Higgins; to my agent, Chris Calhoun; and to my editor, Bob Weil, and his adjutant, Philip Marino.

Among my regrets, the greatest is that Christopher Hitchens is no longer around to argue over the book. When I asked him for a blurb, he wrote back from the cancer center in Houston where he was undergoing last-chance treatment, "Sling it over . . . I'd be proud." Ten days later he was dead.

Finally, for helping me stave off cosmic torpor, thanks to Jared, Malcolm, Jenny, and, most of all, Jon.

Notes

1. Confronting the Mystery

5 "Time and again": Richard Dawkins, *The God Delusion* (Houghton Mifflin Harcourt, 2006), p. 184.

5 "Maybe the 'inflation'": Dawkins, *God Delusion*, p. 185.

5 "What is it that breathes": Stephen Hawking, *A Brief History of Time* (Bantam Books, 1998), p. 190.

6 "No scientific theory": Henry Margenau and Roy Abraham Varghese, *Cosmos, Bios, Theos* (Open Court, 1992), p. 11.

7 "would have no stability": Arthur O. Lovejoy, *The Great Chain of Being* (Oxford University Press, 1973), p. 168.

9 "a prejudice as deep-rooted": Nicholas Rescher, *The Riddle of Existence* (University Press of America, 1994), p. 17.

10 "Is it not probable": David Hume, *Dialogues Concerning Natural Religion* (Hafner, 1948), p. 60.

12 "All of us": William James, *Some Problems of Philosophy* (Longmans, Green, 1911), p. 46.

2. Philosophical Tour d'Horizon

17 "the darkest in all": James, *Some Problems of Philosophy*, p. 46.

17 "tear the individual's": A. C. B. Lovell, *The Individual and the Universe* (Mentor, 1961), p. 125.

17 "constitutes one of": Lovejoy, *Great Chain of Being*, p. 329.

17 "grazed by its hidden power": Martin Heidegger, *An Introduction to Metaphysics* (Yale University Press, 1959), p. 1.

18 "The lower a man": Arthur Schopenhauer, *The World as Will and Representation* (Dover, 1966), vol. 2, p. 161.

19 "It has always": John Colapinto, "The Interpreter," *The New Yorker*, April 16, 2007, p. 125.

20 "This principle having": Gottfried Wilhelm Leibniz, *Philosophical Papers and Letters*, ed. Leroy E. Loemker (University of Chicago Press, 1956), vol. 2, p. 1038.

21 "Whatever we can": Hume, *Dialogues Concerning Natural Religion*, p. 58.

22 "the balance wheel": Schopenhauer, *World as Will*, p. 171.

22 "fools": Ibid., p. 185.

22 "the main function": Friedrich Schelling, quoted in *The Oxford Companion to Philosophy*, ed. Ted Honderich (Oxford University Press, 1995), p. 800.

22 "the vanishing of being": G. F. W. Hegel, *The Logic of Hegel*, trans. William Wallace (Clarendon Press, 1892), p. 167.

22 "spice-seller's explanations": Søren Kierkegaard, *Concluding Unscientific Postscript*, trans. David F. Swenson and Walter Lowrie (Princeton University Press, 1968), p. 104.

22 "I want to know": Henri Bergson, *Creative Evolution*, trans. A. Mitchell (Modern Library, 1944), pp. 299–301.

22 "deepest": Martin Heidegger, *Introduction to Metaphysics*, p. 2.

23 "being able to ask": Ibid., p. 206.

23 "Aesthetically, the miracle": Ludwig Wittgenstein, *Notebooks, 1914–1916*, trans. G. E. M. Anscombe (Harper Torchbook, 1969), p. 86.

24 "the ultimate ontological": Quoted in A. J. Ayer, *The Meaning of Life* (Scribner, 1990), p. 23.

24 "Supposing": A. J. Ayer, *The Meaning of Life* (Scribner, 1990), p. 24.

24 "incredibly shallow": Quoted in Ray Monk, *Ludwig Wittgenstein* (Free Press, 1990), p. 543.

25 "I should say": Quoted in John Hick, *The Existence of God* (Collier, 1964), p. 175.

25 "to that primordial": Address of Pope Pius XII to the Pontifical Academy of the Sciences, November 22, 1951.

25 "scientists who effectively": Quoted in F. David Peat, *Infinite Potential* (Perseus, 1996), p. 145.

25 "Some younger scientists": Quoted in Hans Küng, *Credo* (Doubleday, 1993), p. 17.

25 "the notion of a beginning" Quoted in Helge Kragh, *Cosmology and Controversy* (Princeton University Press, 1996), p. 46.

25 "a party girl": quoted in Jane Gregory, *Fred Hoyle's Universe* (Oxford University Press, 2005), p. 39.

27 "If the universe": Quoted in Margenau and Varghese, *Cosmos, Bios, Theos*, p. 5.

28 "sucking things into": Robert Nozick, *Philosophical Explanations* (Harvard University Press, 1981), p. 123.

28 "Someone who proposes": Ibid., p. 116.

29 "in philosophy": Marcel Proust, *In Search of Lost Time*, trans. D. J. Enright et al. (Modern Library, 2003), vol. 3, p. 325.

29 "were asking one": Timothy Williamson, in *Proceedings of the 2004 St. Andrews Conference on Realism and Truth*, ed. P. Greenough and M. Lynch (Oxford University Press, forthcoming).

30 "from nothing to": James, *Some Problems of Philosophy*, p. 40.

31 "philosophy, like the overture": Schopenhauer, *World as Will*, p. 171.

31 "When you have understood": Quoted in John Updike, *Hugging the Shore* (Vintage Books, 1984), p. 601.

31 "choked with rage": Jean-Paul Sartre, *Nausea*, trans. Lloyd Alexander (New Directions, 1964), p. 134.

32 "a vacuum is a hell": Quoted in John D. Barrow, *New Theories of Everything* (Oxford University Press, 2007), p. 93.

32 "looked around the ring": John Updike, *Bech* (Fawcett, 1965), p. 131.

32 "He believed, if": Ibid., p. 175.

34 "I believe in Spinoza's": Quoted in *Einstein for the 21st Century*, ed. Peter Galison et al. (Princeton University Press, 2008), p. 37.

35 "I see it": Quoted in Joseph W. Dauben, *Georg Cantor* (Harvard University Press, 1979), p. 55.

35 "He who has not": Quoted in *The Encyclopedia of Philosophy*, ed. Paul Edwards (Macmillan, 1967) vol. 8, p. 302.

Interlude: The Arithmetic of Nothingness

39 "Opposites": P. W. Atkins, *The Creation* (W. H. Freeman, 1981), p. 111.

40 "a little speck of": David K. Lewis, *Parts of Classes* (Blackwell, 1991), p. 13.

3. A Brief History of Nothing

41 "*nothing* (n.)": *Webster's New World Dictionary of the American Language*, ed. David B. Guralnik (William Collins, 1976), p. 973.

41 "simpler and easier": Gottfried Wilhelm Leibniz, *Philosophical Papers and Letters*, ed. Leroy E. Loemker (University of Chicago Press, 1956), vol. 2, p. 1038.

42 "The less anything is": *The Works of John Donne*, vol. 6, ed. Henry Alford (John W. Parker, 1839), p. 155.

42　"that which God": Quoted in John Updike, *Picked-Up Pieces* (Fawcett, 1966), p. 97.

43　"Nothingness haunts being": Jean-Paul Sartre, *Being and Nothingness*, trans. Hazel E. Barnes (Philosophical Library, 1956), p. 11.

43　"fullness of being": Ibid., p. 9.

43　*"Anxiety reveals"*: Martin Heidegger, *Basic Writings*, ed. David Farrell Krell (Harper-Collins, 1993), p. 101.

43　"Nothing is neither": Quoted in John Passmore, *One Hundred Years of Philosophy* (Penguin, 1968), p. 477.

44　"a vacuum force": Nozick, *Philosophical Explanations*, p. 123.

44　"By the time": Myles Burnyeat, review of Nozick's *Philosophical Explanations*, *Times Literary Supplement*, October 15, 1982, p. 1136.

44　"venerable and awesome": Plato, *Theaetetus*, 183e, in *The Collected Dialogues of Plato*, ed. Edith Hamilton et al. (Princeton University Press, 1961), p. 888.

46　"There is just": Bede Rundle, *Why There Is Something Rather Than Nothing* (Oxford University Press, 2006), p. 113.

46　"an embroidery on": Bergson, *Creative Evolution*, p 278.

47　"In order for me": A. Luria, *The Mind of a Mnemonist* (Avon, 1969), pp. 131–132.

49　"Our attempt to": Rundle, *Why There Is Something Rather Than Nothing*, p. 116.

49　"Space is not": Ibid., p. 111.

58　"strictly meaningless": Milton K. Munitz, *The Mystery of Existence* (New York University Press, 1974), p. 149.

58　"strictly a technical convenience": W. V. O. Quine, *Philosophy of Logic* (Prentice-Hall, 1970), p. 53.

59　"a triumph of triviality": Ibid., p. 54.

61　"Among the great things": Quoted in Michael J. Gelb, *How to Think Like Leonardo da Vinci* (Delacorte Press, 1998), p. 25.

4. The Great Rejectionist

63　"If there is": Jim Holt, review of Dawkins's *The God Delusion*, *New York Times Book Review*, October 22, 2006, p. 1.

68　"collapse into non-existence": Quoted in *A Dictionary of Philosophy*, ed. Antony Flew (St. Martin's Press, 1984), p. 80.

71　"Absolute, true": Sir Isaac Newton, "Scholium on Absolute Space and Time," in *Time*, ed. Jonathan Westphal et al. (Hackett Publishing Co., 1993), p. 37.

72　"profoundest": J. J. C. Smart, *Our Place in the Universe* (Blackwell, 1989), p. 178.

76　"the truth always": Richard Feynman, *The Character of a Physical Law* (MIT Press, 1967), p. 171.

76 Suppose you have two equally well-confirmed theories: The example is due to Richard Swinburne.

77 "part of what we mean": Steven Weinberg, *Dreams of a Final Theory* (Pantheon Books, 1993), p. 149.

77 "any God capable": Dawkins, *God Delusion*, p. 176.

5. Finite or Infinite?

85 "How can anything": David Hume, *Dialogues Concerning Natural Religion*, p. 59.

Interlude: Night Thoughts at the Café de Flore

90 "His movement is": Sartre, *Being and Nothingness*, p. 59.

92 "The very same criteria": Richard Swinburne, *Is There a God?* (Oxford University Press, 1996), p. 2.

93 "Why, oh why": Adolf Grünbaum, "Rejoinder to Richard Swinburne's 'Second Reply to Grünbaum,'" *British Journal for the Philosophy of Science*, vol. 56 (2005), p. 930.

93 "a breathtaking piece": Dawkins, *God Delusion*, p. 148.

93 "beyond satire": Ibid., p. 64.

93 "May you rot": Quoted in ibid., p. 89.

6. The Inductive Theist of North Oxford

99 in a 1989 essay: Richard Swinburne, "Argument from the Fine-Tuning of the Universe," in *Physical Cosmology and Philosophy*, ed. John Leslie (Macmillan, 1990), p. 158.

106 "vastly improbable": Richard Swinburne, *The Existence of God* (Oxford University Press, 2004), p. 151.

Interlude: The Supreme Brute Fact

111 "So truly, therefore": Saint Anselm, "Proslogion," in *The Ontological Argument*, ed. Alvin Plantinga (Anchor Books, 1965), p. 5.

111 "a charming joke": Arthur Schopenhauer, "The Fourfold Root of the Principle of Sufficient Reason," in *Ontological Argument*, p. 66.

111 "I remember the precise": *The Basic Writings of Bertrand Russell*, ed. Robert E. Egner et al. (Touchstone, 1961), p. 42.

111 "it is easier": Bertrand Russell, *A History of Western Philosophy* (Touchstone, 1972), p. 586.

111 "infantile": Dawkins, *God Delusion*, p. 80.

113 "A hundred real dollars": Immanuel Kant, *The Critique of Pure Reason*, tr. Norman Kemp Smith (Macmillan, 1929), A599/B627.

113 "lost island": Gaunilo, "On Behalf of the Fool," in *Ontological Argument*, p. 11.

115 "purely rationally": Quoted in Hao Wang, *A Logical Journey* (MIT Press, 1996), p. 105.

115 "tough-minded intellectualism": "Modernizing the Case for God," *Time*, April 5, 1980, p. 66.

116 "It breaches no laws": Alvin Plantinga, "God, Arguments for the Existence of," in *Routledge Encyclopedia of Philosophy*, ed. Edward Craig (Routledge, 1988), vol. 4, p. 88.

118 "a sane and rational": Alvin Plantinga, *The Nature of Necessity* (Oxford University Press, 1974), p. 220.

118 "The premise that": J. L. Mackie, *The Miracle of Theism* (Oxford University Press, 1982), p. 61.

118 "Every philosopher would": Russell, *History of Western Philosophy*, p. 417.

7. The Magus of the Multiverse

120 "Deutsch seems more passionate": Oliver Morton, "The Computable Cosmos of David Deutsch," *American Scholar*, Summer 2000, p. 52.

121 "fairly straightforward": David Deutsch, *The Fabric of Reality* (Penguin, 1997), p. 210.

122 "Arrogant in tone": Jim Holt, review of David Deutsch's *The Fabric of Reality*, *Wall Street Journal*, August 7, 1997.

124 "setting international standards": Morton, "Computational Cosmos," p. 51.

129 "I do not believe": Deutsch, *Fabric of Reality*, p. 17.

129 "It is not enough": Ibid., p. 139.

Interlude: The End of Explanation

133 "Self-subsumption is": Nozick, *Philosophical Explanations*, p. 120.

133 "The ultimate principle": Ibid., p. 134.

134 "a self-subsuming statement": Ibid., p. 138.

134 "for surely never": Swinburne, *Existence of God*, p. 79.

135 "If it is a very deep fact": Nozick, *Philosophical Explanations*, p. 131.

135 "There isn't": Ibid., p. 130.

136 "they exist in independent": Ibid., p. 129.

8. The Ultimate Free Lunch?

138 "The clear light of science": Julian Huxley, *Essays of a Humanist* (Harper & Row, 1969), pp. 107–108.

140 "the most profound development": John Gribbin, *Q Is for Quantum* (Free Press, 1998), p. 311.

141 "Maybe the universe": Quoted in Alex Vilenkin, *Many Worlds in One* (Hill and Wang, 2006), p. 183.

142 "stopped in his tracks": Quoted in John Gribbin, *In the Beginning* (Bullfinch, 1993), p. 249.

142 "In answer to the question": Ed Tryon, "Is the Universe a Vacuum Fluctuation?" *Nature*, vol. 246 (1973), p. 396.

142 "A proposal that": Alan Guth, *The Inflationary Universe* (Addison-Wesley, 1997), p. 273.

145 "A quantum theory": Stephen Hawking, *Black Holes and Baby Universes* (Bantam Books, 1993), p. 61.

146 "with the discovery": Weinberg, *Dreams of a Final Theory*, p. 240.

147 "With his crab-apple cheeks": John Horgan, *The End of Science* (Addison-Wesley, 1996), p. 71.

147 "With or without religion": Steven Weinberg, "A Designer Universe?" *New York Review of Books*, October 21, 1999.

9. Waiting for the Final Theory

160 "supposes that there": Weinberg, *Dreams of a Final Theory*, p. 238.

160 "This may happen": Steven Weinberg, "Can Science Explain Everything? Anything?" *New York Review of Books*, May 31, 2001, p. 50.

161 "The tunneling process": Alex Vilenkin, *Many Worlds in One* (Hill & Wang, 2006), p. 204.

161 "What is it that breathes": Stephen Hawking, *A Brief History of Time* (Bantam, 1998), p. 190.

162 "The whole modern conception": Ludwig Wittgenstein, *Tractatus Logico-Philosophicus*, trans. D. F. Pears and B. F. McGuinness (Humanities, 1961), 6.371.

Interlude: A Word on Many Worlds

164 "a trillion trillion other universes": Swinburne, *Is There a God?* p. 68.

164 "there is not a shred": Martin Gardner, *Are Universes Thicker Than Blackberries?* (W. W. Norton, 2004), p. 9.

164 "invoking an infinity": Paul Davies, "A Brief History of the Multiverse,": op-ed, *New York Times*, April 12, 2003.

168 "Surely the conjecture": Gardner, *Are Universes Thicker*, p. 9.

168 "there is no reason": Davies, "A Brief History."

169 "The many-worlds": Leonard Susskind, *The Cosmic Landscape* (Little, Brown, 2005), p. 317.

170 "It is probable": Quoted in Paul Davies, *The Mind of God* (Touchstone, 1992), p. 140.

10. Platonic Reflections

172 "there exists, independently": Alain Connes and Jean-Pierre Changeux, *Conversations on Mind, Matter, and Mathematics* (Oxford University Press, 1995), p. 26.

172 "Mathematicians should have": Quoted in Thomas Tymoczko, *New Directions in the Philosophy of Mathematics* (Princeton University Press, 1998), p. 26.

172 "We do have something": Kurt Gödel, "What Is Cantor's Continuum Problem?" in *Philosophy of Mathematics*, ed. Paul Benacerraf and Hilary Putnam (Cambridge University Press, 1983), p. 484.

172 "unreasonable effectiveness": Eugene Wigner, "The Unreasonable Effectiveness of Mathematics in the Natural Sciences," in *Communications in Pure and Applied Mathematics*, vol. 13, no. 1 (February 1960), pp. 1–14.

172 "You can recognize": Richard Feynman, *The Character of a Physical Law* (MIT Press, 1967), p. 171.

172 "book of nature": Galileo, *Saggiatore*, Opere VI, quoted in *The Penguin Book of Curious and Interesting Mathematics*, ed. David Wells (Penguin Books, 1997), p. 151.

173 "God is a mathematician": Quoted in John D. Barrow, *Pi in the Sky* (Oxford University Press, 1992), p. 292.

174 "I imagine that": Roger Penrose, *The Emperor's New Mind* (Oxford University Press, 1989), p. 428.

174 "His argument is": Quoted in Matt Ridley, *Francis Crick* (Eminent Lives, 2006), p. 197.

180 "To me the world": Roger Penrose, *Shadows of the Mind* (Oxford University Press, 1994), p. 417.

180 "eternally existing": Ibid., p. 428.

180 "profound and timeless": Penrose, *Emperor's New Mind*, p. 95.

180 "ancient and honorable": W. D. Hart, *The Evolution of Logic* (Cambridge University Press, 2010), p. 277.

181 " 'Imaginary universes' are": G. H. Hardy, *A Mathematician's Apology* (Cambridge University Press, 1940), p. 135.

182 "the essence of mathematics": Quoted in Loren Graham and Jean-Michel Kantor, *Naming Infinity* (Harvard University Press, 2009), p. 199.

182 "The elements of this multiverse": Max Tegmark, "Parallel Universes," *Scientific American*, May 2003, p. 50.

182 "have an eerily real feel": Ibid., p. 49.

183 "It's just a feeling": Quoted in Davies, *Mind of God*, p. 145.

183 "Rightly viewed": Bertrand Russell, *Mysticism and Logic* (Doubleday, 1957), p. 57.

183 "largely nonsense": *Basic Writings of Bertrand Russell*, p. 255.

183 "To be is": Willard Van Orman Quine, *From a Logical Point of View* (Harper Torchbooks, 1953), p. 15.

184 "We have the same": Hart, *Evolution of Logic*, p. 279.

185 "Avaunt! You": Bertrand Russell, *Nightmares of Eminent Persons* (Touchstone, 1955), p. 46.

Interlude: It from Bit

187 "What reason": Quoted in Marc Lange, *Introduction to the Philosophy of Physics* (Blackwell, 2002), p. 168.

188 "Kick at the rock": Richard Wilbur, "Epistemology," in *New and Collected Poems* (Harcourt Brace Jovanovich, 1988), p. 288.

188 "in language": Quoted in Jonathan Culler, *Saussure* (Fontana, 1985), p. 18.

189 "all mathematical structures": Tegmark, "Parallel Universes," p. 50.

189 "Our knowledge": Arthur Eddington, *The Nature of the Physical World* (Cambridge University Press, 1928), p. 258.

190 "at the most basic ontological": Frank Tipler, *The Physics of Immortality* (Anchor Books, 1997), p. 209.

191 "The subjective features": Thomas Nagel, *The View from Nowhere* (Oxford University Press, 1986), p. 15.

191 "frankly, quite crazy": John R. Searle, *Mind* (Oxford University Press, 2004), p. 217.

192 "Postulating special inner qualities": Daniel Dennett, *Consciousness Explained* (Little, Brown, 1991), p. 450.

192 "The world just *isn't*": Nagel, *View from Nowhere*, p. 15.

193 "What has structure": T. L. S. Sprigge, *Theories of Existence* (Penguin, 1984), p. 156.

193 "rather like the kind": T. L. S. Sprigge, "Panpsychism," in *Routledge Encyclopedia of Philosophy*, ed. Edward Craig (Routledge, 1988), vol. 7, p. 196.

193 "the stuff of the world": Eddington, *Nature of the Physical World*, p. 276.

194 "Experience is information": David Chalmers, *The Conscious Mind* (Oxford University Press, 1996), p. 305.

195 "How can many consciousnesses": William James, *Writings, 1902–1910* (Library of America, 1988), p. 723.

195 "Take a sentence": William James, *Principles of Psychology* (Dover, 1950), vol. 1, p. 160.

196 "the unity of a single mind": Penrose, *Shadows of the Mind*, p. 372.

196 "I think that something of this nature": Roger Penrose, *The Large, the Small, and the Human Mind* (Cambridge University Press, 1997), p. 175.

196 "absurd": John R. Searle, *The Mystery of Consciousness* (New York Review of Books, 1997), p. 156.

11. "The Ethical Requiredness of There Being Something"

198 "the most interesting": Larry Kaufman, www.hostagechess.com.

200 "However complex the object": James, *Principles of Psychology*, vol. 1, p. 276.

205 "The notion": Mackie, *Miracle of Theism*, p. 232.

213 "the noblest and most": Russell, *History of Western Philosophy*, p. 569.

214 "Each act of cruelty": Ibid., p. 580.

214 "overwhelming bleakness": Interview with Father Robert E. Lauder, *Commonweal*, April 15, 2010.

Interlude: An Hegelian in Paris

218 "Pure Being makes": Hegel, *Logic of Hegel*, p. 135.

218 "simple and indeterminate": Ibid.

218 "This mere Being": Ibid., p. 137.

218 "is just Nothing": Ibid.

218 "No great expenditure": Ibid., p. 140.

218 "an unsteady unrest": Ibid.

219 "The worse your logic": Russell, *History of Western Philosophy*, p. 746.

219 "an ontological proof of": Arthur Schopenhauer, *On the Fourfold Root of the Principle of Sufficient Reason*, trans. Mme. Karl Hillebrand (George Bell and Sons, 1897), p. 13.

219 "The Idea, as unity": Hegel, *Logic of Hegel*, p. 323.

219 "very obscure": Russell, *History of Western Philosophy*, p. 734.

219 "possess the world": *The Philosophy of Jean-Paul Sartre*, ed. Robert Denoon Cumming (Modern Library, 1965), p. 331.

12. The Last Word from All Souls

222 "Why Anything? Why This?" Derek Parfit, *London Review of Books*, January 22, 1998, and February 5, 1998. All Parfit quotations in the chapter are from this essay unless otherwise noted.

223 "The truth is very different": Derek Parfit, *Reasons and Persons* (Oxford University Press, 1984), p. 281.

229 "What interests me": quoted in Steve Pyke, *Philosophers* (Distributed Art Publishing, 1995), p. 43.

229 "a florid antique shop": Christopher Hitchens, *Hitch-22* (Twelve, 2010), p. 103.

231 "In the Ultimate Multiverse": Brian Greene, *The Hidden Reality* (Allen Lane, 2011), p. 296.

231 "Nothingness haunts Being": Sartre, *Being and Nothingness*, p. 11.

13. The World as a Bit of Light Verse

244 "skate upon an intense radiance": John Updike, "The Dogwood Tree," in *Assorted Prose* (Fawcett, 1966), p. 146.

244 "Barth's theology": Updike, preface to *Assorted Prose*, p. viii.

244   "Satanic nothingness": Updike, *Picked-Up Pieces*, p. 99.

246   "cosmic bootstrap": Peter Atkins, *The Creation* (W. H. Freeman, 1981), p. 111.

246   "beset by an embarrassment": Martin Amis, *The War Against Cliché* (Vintage, 2002), p. 384.

252   "blot on nothingness": Updike, *Bech*, p. 131.

14. The Self: Do I Really Exist

253   "How often have I said": Arthur Conan Doyle, *The Sign of the Four* (Spencer Blackett, 1890), p. 93.

254   "Existence precedes essence": Jean-Paul Sartre, "Existentialism Is a Humanism," in *Existentialism from Dostoevsky to Sartre*, ed. Walter Kaufman (Meridian Books, 1956), p. 290.

254   "The purpose is to live": Ivan Goncharov, *Oblomov*, trans. Marian Schwartz (Yale University Press, 2010), p. 254.

255   "the lucky ones": Richard Dawkins, *Unweaving the Rainbow* (Mariner, 2000), p. 1.

255   "There is a general belief": Russell, *History of Western Philosophy*, p. 594.

256   "When I enter": David Hume, *A Treatise of Human Nature* (Oxford University Press, 1888), p. 252.

257   "selves are not": Dennett, *Consciousness Explained*, p. 423.

257   "There simply isn't any 'I' ": Galen Strawson, *Selves: An Essay in Revisionary Metaphysics* (Oxford University Press, 2011), p. 246.

257   "I may understand": Nagel, *View from Nowhere*, p. 42.

258   "I think there is a pain": Charles Dickens, *Hard Times* (Oxford World's Classics, 2008), p. 185.

258   "the most monstrous contradiction": Quoted in *The Oxford Companion to Philosophy*, ed. Ted Honderich (Oxford University Press, 1995), p. 817.

258   "The I is not an object": Wittgenstein, *Notebooks, 1914–1916*, p. 80.

260   "only a conventional name": Quoted in Parfit, *Reasons and Persons*, p. 52

260   "in the most deplorable condition": Hume, *Treatise on Human Nature*, p. 269.

260   "liberating and consoling": Derek Parfit, *Reasons and Persons* (Oxford University Press, 1984), p. 280.

261   "with great hesitation": Nozick, *Philosophical Explanations*, p. 87ff.

261   "All art, religion": Roger Scruton, *Modern Philosophy* (Penguin, 1994), p. 484.

262   "The objective world": Edmund Husserl, *Cartesian Meditations*, trans. Dorion Cairns (Martinus Nijhoff, 1970), p. 26.

263   "How can I": Nagel, *View from Nowhere*, p. 61.

263   "What kind of a fact": Ibid., p. 54.

263   "the world soul in": Ibid., p. 61

263   "the additional thought": Ibid., p. 60.

265   "Amazement that the universe": Ibid., p. 56.

15. Return to Nothingness

266   "It is entirely impossible": Quoted in Paul Edwards, "My Death," in *The Encyclopedia of Philosophy*, ed. Paul Edwards (Macmillan, 1967), vol. 5, p. 416.

267   "Not the least": Quoted in Simon Critchley, *The Book of Dead Philosophers* (Vintage, 2009), p. 176.

267   "If we are to make sense": Thomas Nagel, *Mortal Questions* (Cambridge University Press, 1979), p. 4.

267   "It is not that death": Richard Wollheim, *The Thread of Life* (Yale University Press, 1999), p. 269.

268   "I must confess": Miguel de Unamuno, *Tragic Sense of Life*, trans. Anthony Kerrigan (Princeton University Press, 1972), p. 49.

268   "ownmost death": Mark Johnston, *Surviving Death* (Princeton University Press, 2010), p. 138.

269   "my existence seems": Nagel, *View from Nowhere*, p. 228.

269   "That is all there is": Parfit, *Reasons and Persons*, p. 280.

269   "Awakened to life": Quoted in Scruton, *Modern Philosophy*, p. 378.

270   "We are not at home": Ibid., p. 464.

Epilogue

277   book-chat show: The television program was *Bouillon de Culture*. The Dominican priest was Jacques Arnould, the physicist was Jean Heidmann (who died in 2000), and the Buddhist monk was Matthieu Ricard.

Index